THE STRUCTURE AND
LIFE OF BRYOPHYTES

Biological Sciences

Editor

PROFESSOR A. J. CAIN
MA, D.PHIL
Professor of Zoology
in the University of Liverpool

THE STRUCTURE AND
LIFE OF BRYOPHYTES

E. V. Watson

Senior Lecturer in Botany in the University of Reading

HUTCHINSON UNIVERSITY LIBRARY

LONDON

London Melbourne Sydney Auckland
Wellington Johannesburg Cape Town
and agencies throughout the world

First published 1964
Second edition 1967
Reprinted 1968
Third edition 1971

*This book has been set in Times type, printed in Great Britain
on smooth wove paper by Anchor Press, and
bound by Wm. Brendon, both of Tiptree, Essex*

ISBN 0 09 109300 7 (cased)
0 09 109301 5 (paper)

CONTENTS

74–00111

ACKNOWLEDGMENTS

I should like to acknowledge my debt to the following in respect of material reproduced for certain illustrations: Professor D. J. Carr and the *Australian Journal of Botany* for Figs. 3A and B; Dr S. Williams and the Cambridge University Press for Fig. 3C; the *New Phytologist* for 3D; the late Professor J. Proskauer and the *Bryologist* for 3E; Professor C. T. Ingold and the Clarendon Press for Figs. 10F, J and L; Dr C. Hébant for Figs. 20B and C; the *Annals of Botany* for Fig. 23A; Mme S. Jovet-Ast and the publishers of Boureau's *Traité de Paléobotanique* for Figs. 23D, E and F; the same publishers for Fig. 25B; the late Dr M. F. Neuburg and the publishers of her monograph on Permian Mosses for Fig. 24; Dr R. M. Schuster and *Nova Hedwigia* for Figs. 26A, B and E; Professor R. Grolle and the *Österreiche botanische Zeitschrift* for Fig. 26F.

PREFACE TO THIRD EDITION

It is gratifying to have had the opportunity of preparing this revised edition, but the task of doing so has proved both long and difficult. This is in large measure because the literature that has appeared over the past decade is so extensive that one is faced with the impossibility of doing justice to it. Even so, an attempt has been made to draw the attention of the enquiring student at least to some of the salient publications in each branch of bryophyte study, with the result that rather over two hundred additional references appear in the bibliography.

The general aim of the book remains as before and the plan and principal contents of each chapter are unchanged. The detailed content, however, has been subject to very numerous alterations which I hope will be conceded to be improvements. This is especially true of Chapters 8 to 11 inclusive, whilst Chapter 12 has been almost completely re-written and greatly expanded. As in the first instance, the book is not aimed at the needs of the complete beginner but is intended to provide a lead into the literature for the rather more advanced student. Most of the chapters are in the nature of review essays, in which ample reference to sources must nevertheless allow the intrusion of certain personal judgments and points of view.

I am very conscious of the shortcomings of the text, even in its new, revised form. It can be said with truth, however, that they would have been even graver were it not for the very considerable help I received from a number of fellow bryologists (to whom I wrote for suggestions), especially Mr A. C. Crundwell, Dr P. J. Grubb and Dr E. W. Jones. I am deeply grateful to them all. It is also a pleasure to record my debt to Dr J. G. Duckett for special assistance with parts of Chapter 8, and to Mr P. J. Wanstall for much helpful discussion.

In the new, revised Chapter 12 it has been my aim both to expand the original brief concluding statement, and at the same time to show the ways in which our thoughts on taxonomy and phylogeny are indissolubly linked together. In a sense, too, this chapter draws together some of the threads from earlier chapters. The present-day position is appraised with special reference to taxonomic and distributional studies, and tentatively we look out into the future.

The illustrations, largely because of restricted space available, present almost insuperable problems. One would like to have made them far more numerous, larger, less congested, and to have added here and there some numerical material. Space, however, does not allow this. A few new illustrations are added and the old ones have been provided with fuller legends and are less drastically reduced than before. The 'comparative approach' is as much a part of the plan behind the illustrations as it is inherent in the design of the text.

A principal source of encouragement to me has been the warm reception and continuing sales of the original book in many parts of the world. So too have been the tactful patience of Messrs Hutchinsons (as the months wore on) and the skilled typing of Mrs Ord-Hume without which this new edition would have been even longer delayed. Finally, I would thank Professor V. H. Heywood for many facilities and for his encouragement; and many bryological friends for the interest they have shown.

E. V. WATSON

Reading
March 1971

PREFACE TO FIRST EDITION

The aim of this book is to fill a gap which exists between the full, straightforward treatment of the morphology of bryophytes and the research literature in all its detail and diversity of topics. My hope is that it will enable the university student to see morphological facts from a new angle and at the same time have his interest directed to other branches of bryophyte study.

The chapters in this book take the form of essays on various bryological topics, and, although no claim is made that the treatments given are in any instance exhaustive, a serious attempt has been made to introduce the student to a reasonably copious and representative literature. In the morphological chapters with which the book begins a comparative approach has been adopted, in order that old facts may be seen from a new angle; in order too that the habit of making comparisons may grow in the student of this subject.

It is true that each of the later chapters can provide only a sketchy outline of a big subject. Nevertheless, by dealing with some of these subjects at all one may be able to help the enquirer who is apt to find so little about most of them in the standard texts. I refer to bryophyte ecology, anatomy and physiology; also cytogenetics, geographical distribution and the meagre fossil record of these plants.

To select suitable subjects for the limited number of line drawings that can be included has not been easy. I would have liked more, but perhaps these few may serve to clarify the text in places. I am indebted to those authors and publishing houses that have kindly allowed copies of certain drawings to be made.

It is a great pleasure to record my gratitude to my friend Mr
P. J. Wanstall, who has most kindly read the entire text and suggested a number of improvements. I would also thank Professor
H. Munro Fox, F.R.S., at whose invitation the book was undertaken,
for his helpful criticism in the early stages and keen interest throughout. Finally, I am grateful to Professor T. M. Harris, F.R.S., for his
unfailing interest and help on a number of occasions.

E.V.W.

May 1963

I

INTRODUCTION AND CLASSIFICATION

The structure and life of bryophytes are usually approached by means of the detailed presentation of selected examples from the two groups, mosses and liverworts. In general textbooks of botany a single example from each has often been thought sufficient; in fuller treatments of cryptogams some details of classification are given and more examples taken. An orderly picture of the whole can be built up in this way, but another, quite different, approach is possible and that is the one adopted in this book. Here we are concerned in the different chapters with a review of different aspects of bryophyte study, which on the whole lend themselves to separate treatments, although some overlap is inevitable.

A certain amount of previous experience of the group is assumed. It is the aim of the chapters that follow to enable the student who has already begun a study of these plants to reach out further in many directions, to see something of the scope of the principal branches of bryology and to be led to the more important literature. In the course of the discussions reference will be made to diverse examples. Occasionally the germ of a new idea may come to the surface or a possible future line of investigation suggest itself. We shall start with structure considered mainly in relation to evolutionary problems, and proceed to other topics—vegetative propagation, sex organs, morphogenesis, anatomy and physiology, ecology, bryogeography and some others in later chapters. Some of these are big topics and space will permit only a limited treatment of them; but enough may be set down to indicate something of the breadth of the field which in each instance awaits the attention of the enquiring student.

First of all we may define the limits of the group Bryophyta. It is widely known that bryophytes include the mosses (Musci) and the liverworts (Hepaticae). Many people think of the plant body of a moss as equipped with stem and leaves, whilst that of a liverwort they visualise as thalloid—a flattened, freely branched structure of variable size and shape. It is sometimes forgotten that the leafy liverworts far exceed the thalloid in number of species. Also, the word 'moss' is applied to much that has no connection with bryophytes. Thus reindeer moss is a lichen, club-mosses are pteridophytes (related to ferns and horsetails) and Spanish moss is a highly modified seed plant of the family Bromeliaceae.

The bryophytes are a well-defined and circumscribed group of plants. No recognisable link connects them on the one hand with algae, on the other with pteridophytes. Their evolution remains to a great extent a matter of speculation, and the interrelationships between the different living groups are by no means clearly understood. This arises partly from the fact that some groups, such as the so-called 'leafy liverworts' (Jungermanniales) and the 'true mosses' (Bryidae), are represented by an immense range of forms which are on the whole remarkably uniform, whilst others stand apart, a mere handful of forms or perhaps a few hundred species, in comparative isolation. The genus *Sphagnum* is such an example; *Anthoceros* and its immediate allies furnish another.

Examples from the Bryophyta take their place in elementary books on botany mainly because they illustrate a pattern of life cycle which is peculiarly their own. True, it finds a parallel in certain algae in which there is an alternation between two structurally similar generations; but the bryophyte life-cycle remains unique in that it displays two alternating generations, roughly equivalent in importance if not in size, both highly characteristic in their morphology, yet totally unlike one another. This life-cycle is far removed from the dominance of the haploid generation found in many green algae; it is almost as far removed from the overwhelming dominance of the sporophyte generation that occurs throughout the Pteridophytes. A study of *Pellia* reveals gametophyte dominance, for the green thallus is perennial, independent and freely branching, whilst the sporophyte is of strictly limited growth and duration, and in a fairly complete sense parasitic. *Buxbaumia,* a genus of partly saprophytic mosses, however, conveys quite another impression, with its sporophyte inordinately large beside the insignificant gametophyte (Fig. 13A). So there is some latitude in this matter of the emphasis that rests on each of the two generations; but the salient facts learnt regarding the life-cycles of the liverwort *Pellia* and the moss *Funaria* are not only very similar but also hold good throughout the bryophytes, with only rather unimportant variations.

We may now outline those features of the bryophyte life-cycle which are true of the group as a whole. A haploid spore germinates to produce a stage, commonly filamentous but sometimes ovoid, spherical or plate-like, which precedes the formation of the leafy shoots or gametophores. The latter are typically green, live for anything from a few months to several years, and bear the sex organs, antheridia and archegonia. In every instance, a motile male gamete from an antheridium having fertilised the egg, or oosphere, which lies at the base of the flask-shaped archegonium, a zygote is formed and the sporophyte generation supervenes. In all but the simplest cases this diploid phase early develops polarity and differentiates into a basal foot, a stalk-like seta and a distal spore-producing organ, the capsule. When the sporogenous tissue (archesporium) has completed the necessary number of cell divisions leading to the formation of spore mother cells the latter undergo each a meiotic, or reduction division, and tetrads of spores are formed which normally separate before discharge. Dehiscence and spore dispersal frequently entail the formation and functioning of structures, such as the elaters of liverworts and the peristome of mosses, which belong peculiarly to the Bryophyta. It may be noted, moreover, that both archegonium and antheridium are seen in a more complete and elaborate form here than in any other group of plants.

The structure of each generation and the mode of fertilisation have imposed certain restrictions on the size of bryophytes. The largest normal erect-growing leafy stems are probably those of the larger species of the Australasian genus *Dawsonia* (40–70 cm); but Martin records[284] a remarkable example (seen by him in New Zealand) of a stem of *Polytrichum commune* attaining a length of six feet (150 cm) under water. Again, great length is attained by the underwater shoots of an aquatic moss such as *Fontinalis antipyretica* and by the pendent masses of many tropical epiphytes; but bryophytes as a whole are small and many are indeed microscopic. Few mosses have leaves exceeding 10 mm in length, whilst those of leafy liverworts are much smaller. The capsules of mosses are rarely above 6 mm long; those of liverworts are usually quite minute. Yet, small though they are, bryophytes can be important ecologically. Only seldom, as in some kinds of bog, arctic tundra and tropical mossy forest, do they compose the main element in the vegetation, but their value as indicator species is coming increasingly to be recognised.

Following his grounding in *Pellia* and *Funaria,* the student usually proceeds to a study of further examples, with the emphasis mainly on comparative morphology. Prepared slides have often been the basis of such a study, for it is only in the permanent preparation, made from material that has been embedded and microtomed, that many of the relevant details may be seen. Such studies are, of course, invaluable as a training in interpretative morphology. It is, however, essential that they be supplemented by a generous allowance of time devoted to the examination and dissection of freshly gathered plants. Thereby, the student can begin to see how the microscopic features are related to the life of the organism, and by working with a flora he can gradually acquire a knowledge of bryophyte taxonomy.

If a student is to attain proficiency in this study, he must be prepared to spend time in the field, and even more in the close microscopic study of the plants that he finds. In my view he will do well to move away at an early stage from too implicit a reliance on keys, and try to gain some real insight into the facts and ideas behind the scheme of classification. Then he will fairly soon be able to refer a proportion of the plants found directly to their genus, or to one of a small group of genera. In my experience every competent systematist —in whatever group of organisms—possesses this faculty for seeing the range of form exhibited by his group as a whole, and it is useless to suppose that one can proceed very far without it. The prerequisites are a keen eye for form and plenty of application. A real sense of power results when one becomes aware, for the first time, of having some true command of a group, such as mosses or liverworts, or for that matter the flowering plants or the birds of Britain. Perhaps there are two criteria by which we can judge that this time has come; first, when one can recognise at sight the bulk of what one finds; and secondly, when one can walk anywhere in the land with the knowledge that nothing—or almost nothing—that one finds will be wholly unfamiliar.

Two useful aids for the student of bryophytes are the herbarium and the collection of species alive in cultivation. The former is indispensable for serious taxonomic study, and most keen students will wish to build up their own reference collection. A range of species in cultivation, in unglazed pans in the cool greenhouse, is also an asset, especially where one may wish to see changes that take place in a particular plant throughout the year. Thallose liverworts are the easiest and most satisfactory bryophytes to grow by this method, whilst many mosses seem to suffer because one cannot replicate the conditions that they found in nature. Useful instructions are given by Richards[361] and those bryologists who have the facilities should experiment for themselves. By such methods one can add greatly to one's knowledge of the species studied; especially as regards their range of structural variation and their growth rate.

Equipped with some all-round knowledge, one can go on to specialise, on a chosen group or species, as an ecologist or in some other direction. The morphologist will be encouraged by Fulford's remark[129] that of all the known genera of the great order Jungermanniales (leafy liverworts) only some 5% had been fully investigated morphologically (by 1948). The ecologist will find that comprehensive autecological studies have covered a far smaller proportion of the known species. The taxonomic specialist will find genus after genus of bryophytes awaiting a critical revision—even on a regionally limited basis. Finally, of course, there are those who turn to bryophytes only for some special advantage that they may offer for a physiological or genetical experiment.

Mosses and liverworts have long challenged taxonomists to devise a classification which will be at once reasonably natural and clearly workable. This challenge has not been easy to meet. It will be convenient to begin by considering the liverworts. The group is currently classified along the following lines (Schuster[384]):

Class I   Hepaticae
    Subclass A   Jungermanniae
        Order 1   Takakiales (single genus, *Takakia*)
        Order 2   Calobryales (single genus, *Haplomitrium*)
        Order 3   Jungermanniales (a very large and fairly diverse order—the 'leafy liverworts')
        Order 4   Metzgeriales (includes mainly thalloid liverworts, e.g. *Pellia* and allies)
    Subclass B   Marchantiae
        Order 5   Sphaerocarpales (small group)
        Order 6   Monocleales (single genus, *Monoclea*)
        Order 7   Marchantiales (includes entirely thalloid liverworts among which the thallus attains maximal complexity)
Class II   Anthocerotae
    Order 8   Anthocerotales

This classification supersedes earlier ones in most of which the liverworts were grouped into six orders. The two additional orders are the Takakiales and Monocleales. The genus *Takakia* (Fig. 26, p. 185) represents a recent discovery and *Monoclea* has always been something of an enigma (cf. Fig. 3 p. 33). The many remarkable features of the Anthocerotales seem to justify the placing of this order in a subclass apart. The anomalous character of this group has long been recognised (Leitgeb,[257] Cavers[74]) and as early as 1899 Howe[195] elevated it to class status under the name Anthocerotae.

Of the remaining orders, it will be noted that two (Calobryales, Sphaerocarpales) are small, two are of medium size (Marchantiales, Metzgeriales) and one (Jungermanniales) is very large. Schuster recognises eight families within the order Metzgeriales, eleven within Marchantiales and no fewer than thirty-six in the Jungermanniales. The criteria for the separation of one family from another are not always as sharply defined or as easily appreciated as one could wish. Perhaps partly for this reason, the family concept does not have among liverworts the prominent place it has always had in the taxonomy of flowering plants.

It will be convenient now to turn to the question of the arrangement of the principal groups of mosses. Nobody has ever found a plant that is manifestly intermediate between liverworts and mosses, so that (leaving aside Anthocerotae) these two classes stand as separate major entities within the Bryophyta. They differ in rhizoid structure, in manner of development of sex organs, in the prevailing mode of growth and cell structure of their leaves; in liverworts the seta develops rapidly at a late stage, and the capsule typically contains elaters in addition to the spores; whilst in mosses the seta develops slowly over a long period, and the capsule contains no elaters but normally has a characteristic ring of peristome teeth; and the moss sporophyte has a green phase not paralleled in liverworts outside the Anthocerotales. These and other features amply separate the two great groups.

Some 14,000 species of moss are known and the great majority are sufficiently alike in structure to create a real difficulty for the taxonomist, especially in the matter of fixing the limits of families and genera. A few groups of mosses, however, stand apart in isolation. This fact is reflected in all classifications of mosses that have been proposed. The groups in question are (1) the bog-mosses referable to the genus *Sphagnum* and (2) a group comprising the small, dark-coloured and mainly rock-dwelling mosses of the genus *Andreaea,* together with one other genus, *Neuroloma.* Whether the genus *Sphagnum,* with its many species distributed throughout the world, is quite so isolated as most bryologists have believed it to be has been questioned;[137] but it certainly appears so in its biology, its peculiar gross morphology and fine leaf anatomy, and in other gametophyte characters, besides important features of the sporophyte. *Andreaea,* represented in Britain by several species, and widely distributed in suitable parts of every continent, has nothing very remarkable about its leafy shoots, but shows distinctive features in protonema and sex organs. Like *Sphagnum*, it has the capsule raised on a gametophyte stalk (pseudopodium); and it displays an almost unique mode of dehiscence, the capsule splitting along four lines and releasing the spores through the gaping slits so formed.

To a lesser extent, two other groups of mosses stand apart. These are (1) the curious group represented in Britain by *Buxbaumia* and *Diphyscium*, and (2) the group typified in its highest development (in the northern hemisphere) by *Polytrichum*, but with an interesting southern hemisphere counterpart in *Dawsonia*, which is probably less closely related than it at first appears.

The classification given by Reimers[357] in Engler's Syllabus takes account of these facts and admits five sub-classes of mosses. The system is shown here in outline:

Class	Musci	
Subclass	Sphagnidae	
Order	Sphagnales	1 Family
Subclass	Andreaeidae	
Order	Andreaeales	1 Family
Subclass	Bryidae	
12 Orders, as follows:		
	Archidiales	1 Family
	Dicranales	7 Families
	Fissidentales	1 Family
	Pottiales	3 Families
	Grimmiales	1 Family
	Funariales	6 Families
	Schistostegales	1 Family
	Tetraphidales	1 Family
	Eubryales	16 Families
	Isobryales	23 Families
	Hookeriales	6 Families
	Hypnobryales	12 Families
Subclass	Buxbaumiidae	
Order	Buxbaumiales	2 Families
Subclass	Polytrichidae	
Order	Polytrichales	1 Family
	Dawsoniales	1 Family

The differences are indeed slight between most modern systems of classification and they are concerned mainly with rank and terminology, only rarely with the position, or even the existence of a given order, of mosses. This does not mean that there is unanimity on the question of the probable course of evolution in the Musci, but rather that most contemporary bryologists are agreed to follow in the main the system of arrangement which came into being shorty after the turn of the century and owed its foundation to men such as Fleischer[122] and Brotherus,[50] who were acquainted with mosses on a world scale. If one compares this arrangement with those set forth in Dixon's *Student's handbook of British mosses* or in Grout's *Moss flora of North America*,[157] the most substantial difference seems to lie in the fact that earlier authors took no account of the ordinal rank; thus Dixon[107] designated Sphagnales, Andreaeales and Bryales subclasses and within the Bryales recognised twenty-five 'orders' in Britain. These 'orders' carried the family suffix -*aceae* (e.g. Hypnaceae) and did not correspond with the orders of today.

A dilemma which has always bothered the bryologist concerned with moss classification is that of the rival claims of gametophyte and sporophyte to provide evidence on which he may base his conclusions. Dixon himself, at a later date,[108] expressed this clearly when he presented a more modern scheme of classification as his contribution to Verdoorn's *Manual of bryology*. It is safe to say that this dilemma has never been completely overcome but at an earlier date the emphasis tended to be placed very heavily on the sporophyte and especially on the peristome. More recently gametophyte characters have been given more weight. The distinction was made early between the prevailingly erect or ascending 'acrocarps', with their terminal archegonia and sporophytes, and the commonly prostrate, freely branched 'pleurocarps' with archegonia and sporophytes borne laterally. It is still useful on occasion, although its validity was seriously questioned long ago.[74] In the system which we are adopting almost all the 'pleurocarps' are found in three orders: Isobryales, Hookeriales and Hypnobryales.

The student who consults a larger textbook or a flora must be prepared to meet minor differences in the classifications of both liverworts and mosses from the ones here given in outline. An important check-list of European and North American hepatics was published in 1939 by Buch, Evans and Verdoorn[58] and some contemporary authors follow this with various modifications. One such author is Arnell.[16] In large measure Jones based his Annotated list of British Hepatics on Müller[300] whose *Die Lebermoose Europas* is by far the most authoritative work on the liverworts of Europe. Completed in 1957, it naturally shows many differences, both in classification and nomenclature, from the system to which British bryologists have grown accustomed in Macvicar's Handbook[277] (the most recent edition of which appeared thirty years earlier). Müller differs from Schuster chiefly in the precise content of some of the families. For mosses, the check-list of Richards and Wallace,[363] and Nyholm,[309] in her recently completed *Moss flora of Fennoscandia*, will each show minor differences from Reimers. Again, in some textbooks[403] the student will come across Sphagnobrya, Andreaeobrya and Eubrya for the three subclasses of Musci.

In the classification of every group of plants and animals stability remains an unattainable ideal and the user must ever be prepared for change. In the Bryophyta it is perhaps remarkable that there is a fair measure of agreement between the various proposed classifications of both liverworts and mosses. There is little doubt that the two considerations which militate most strongly against any real certainty of conclusions are (1) the comparative scarcity of fossil bryophytes and (2) the difficulty of according due weight to the often conflicting evidence afforded by the two generations—gametophyte and sporophyte.

GAMETOPHYTE STRUCTURE OF
THALLOID LIVERWORTS

In this chapter we consider thalloid structure among the gameto-phytes of liverworts from a comparative point of view. Accounts of the thallus of *Pellia* appear in all elementary textbooks of botany and these are supplemented in bryophyte textbooks[320,403] by those of further selected types. Usually included are *Riccia, Marchantia* or a close ally such as *Preissia,* and *Anthoceros.* Sometimes *Riccardia* finds a place. In few cases, other than the masterly morphological treatment of Cavers,[74] and the more functional or 'biological' account of Goebel,[148] has the comparative approach been adopted. In 1966, however, there appeared *The Hepaticae and Anthocerotae of North America* (vol. 1) by R. M. Schuster,[384] which adopts the comparative approach on a scale far more generous than is possible here and represents an almost inexhaustible treasure house of information.

At first glance, a sharp boundary seems to separate thalloid from 'leafy' construction. If we compare *Pellia* with a liverwort such as *Cephalozia* we feel reassured that this is so. Liverworts with a ribbon-shaped, dichotomising plant body, after the manner of many sea-weeds, are unquestionably thalloid; those which display differentiation into cylindrical axis and dorsiventral appendages disposed in some regular manner upon it are equally clearly leafy. Few would quibble about the appendages being termed leaves. The difficulty arises, however, with a gametophyte which is in a general way thalloid but is so strongly and regularly lobed that the lobes or 'wings' so formed might be construed by some morphologists as 'leaves'. Such a line of thought has resulted in the claim by Proskauer[345] that the simple and delicate (but sometimes strongly lobed or subdivided) gametophyte of *Sphaerocarpos* is 'fundamentally leafy'.

It is a claim which, whatever its merits, does not alter the fact that *Sphaerocarpos can* display the simplest kind of thalloid construction we can imagine. In this form it has been in cultivation for twenty years at Reading and it may perhaps be taken as our first illustrative example in this chapter. From it we may go on to consider other thalli of more complex form.

The delicate, variously lobed plate of tissue which constitutes the plant body of *Sphaerocarpos michelii* on these occasions is so plentifully covered by pear-shaped 'involucres' that little else is at first sight visible. When these are removed it is seen to be a very frail structure, quite without internal differentiation. Simple rhizoids emerge from the central 'cushion' of tissue, which is several cells thick in contrast with the wings which are composed of a single layer. Male plants can be as little as 1 mm in diameter. The older morphologists considered that such an undifferentiated cell plate could not be far removed from a hypothetical algal ancestor. This thought so attracted the early evolutionary botanist Lotsy[268] that he wrote of an imaginary *'Sphaero-Riccia'* which would serve as an ancestral bryophyte (if it could be found), combining the simple gametophyte of *Sphaerocarpos* with the remarkably undifferentiated sporophyte of *Riccia*.

New cells come from groups of apical initials, much as they do in *Marchantia*. Sex organs arise in sequence from the apex, associated with mucilage papillae ('hairs'). It is instructive to dissect out developing archegonia and to observe how, at an early stage, the pear-shaped involucre grows up to invest each one individually. At maturity they come to be specialised features of an otherwise very simple gametophyte. Despite its small size and apparent simplicity, the gametophyte of *Sphaerocarpos* is not commonly cited by contemporary morphologists as an 'evolutionary starting-point'. Some reasons for this will become apparent at a later stage.

The widely studied genus *Pellia,* of the Metzgeriales, may be cited as our second example. It is truly thalloid, with no debatably 'leafy' states such as some exotic species of *Sphaerocarpos* display. A vertical section shows that abundantly chlorophyllose cells appear only in the superficial layers of dorsal and ventral surfaces (Fig. 2A). Otherwise there is little tissue differentiation and, since the tissues are without air spaces, the thallus appears deep green and translucent. The rhizoids are simple and uniform. Antheridia, followed by groups of archegonia, are elaborated from cells cut off in sequence from the large, four-sided apical cell. They mature along the upper surface of the thallus midrib. Old, discoloured antheridial 'craters' are a feature of the mature thallus. The form of the involucre varies from species to species. Although the plant is so much larger than in *Sphaerocarpos,* this is a simple type of gametophyte and the two genera share numerous features in common. In an earlier arrangement, indeed, Schiffner[376] placed 'Sphaerocarpoid' liverworts in the same order as *Pellia.* If gametophyte evidence alone were available they might still be so placed today.

No such similarity to any other known group is shown by the thallus of *Preissia quadrata,* which typifies the 'higher' Marchantiales. It is far more complex than either of the two plants considered so far and quite different biologically. For here the thallus is not merely many cells thick (approximately thirty cells in the midrib, ten cells in the wings), but it is sharply differentiated into a narrow, chlorophyll-rich upper (dorsal) region and a colourless, lower (ventral) region (Fig. 2B). The ventral surface of the thallus bears two kinds of rhizoids and rows of ventral scales. The upper surface is divided into 'areolae', each of which marks the limits of an underlying air chamber, the entry to which is by way of a centrally placed, barrel-shaped air pore. The walls of each chamber are colourless cells, and from the floor arise short chains of ovoid or subspherical cells extremely rich in chloroplasts. Beneath this well aerated and highly developed photosynthetic system the layers of colourless cells extending to the ventral surface are themselves differentiated (in the midrib) to include storage cells of several kinds, whilst here and there a thick-walled fibre may be seen. Often, on either side of the middle line there will be extensive patches of cells with purple walls, many of them containing the hyphae of a mycorrhizal fungus.

In short, here is a thallus far more complex in structure than anything found elsewhere among liverworts, and similar only to that of its own immediate allies such as *Marchantia* and *Conocephalum.* It would seem, moreover, to be fitted biologically to carry on efficient absorption from the substratum by means of abundant rhizoids and, at the same time, preserve the dorsal layers of the thallus for a functional life akin to that of the green leaf among vascular plants. Finally, the ventral scales provide some protection for the delicate apical tissues. The sex organs, borne up on stalked receptacles (the male smaller and more shortly stalked), serve only to enhance the impression that here is the climax of thalloid evolution among liverworts.

The well-known thallus of *Anthoceros* is superficially not unlike that of *Pellia,* for the rosettes, or irregularly shaped gametophytes, are typically dark green and without obvious gloss. The outline of the thallus is wavy, the surface sometimes crimped in a characteristic way. It lacks the greatly thickened midrib of *Pellia* but growth is from an apical cell, rather as in the commoner genus; and only rhizoids with smooth walls occur. In some species of *Anthoceros* (but not in *A. laevis*) there are extensive cavities internally, but these are filled with mucilage, not air. Some are occupied by the blue-green alga *Nostoc*, and in 1967 Ridgway[364] (cf. Peirce[327]) produced evidence of nitrogen fixation in these colonies.

An important feature of *Anthoceros* is the single large chloroplast which ordinarily occurs in each green cell of the gametophyte. This situation is unknown in all other orders of bryophytes (where numerous small discoid chloroplasts prevail). The single chloroplast, and the characteristic 'pyrenoid' normally associated with it, are akin to those of green algae and of great evolutionary interest. Parihar[320] (p. 129) gives a good short account of them, and points out how our knowledge of the relationship between these chloroplasts and the normal, much smaller kind has been advanced by means of the electron microscope. Very recently Burr[62] has detected, within the Anthocerotae, a trend away from the algal type of chloroplast.

A second remarkable feature is that both kinds of sex organ develop within, rather than above the surface tissues. This makes the mature archegonia appear fully immersed, their venters and necks confluent with surrounding vegetative cells; whilst the antheridia are found, several together, on the floor of cavities that are roofed over at first but are ultimately broken through to form conspicuous antheridial 'craters' whose orange contents are at times very obvious indeed. The full justification for the separation of the Anthocerotales from all other groups of liverworts can be appreciated only after taking into account the unique features of the sporophyte; but the gametophyte is surely remarkable enough to suggest that *Anthoceros* is not very closely connected with the other three forms that we have examined.

We may now enquire how far these four examples represent the range of thalloid gametophyte structure among liverworts and what light is thrown upon them by a survey of the immediate allies of each. This will be instructive, both from the biological and the evolutionary standpoints; but it must be emphasised that evolutionary conclusions must rest also on facts drawn from the sporophytes.

In every order except the Marchantiales we see a close relationship between forms that are indisputably thalloid and others which show what may be called a 'quasi-leafy' condition. It has already been pointed out that such a situation exists within the genus *Sphaerocarpos*. Allied to *Sphaerocarpos* is the aquatic genus, *Riella* with its unique combination of slender branching axis and broad, wavy wing that is sometimes subdivided into leaf-like entities. *Pellia* is not far removed from forms which were interpreted by the earlier morphologists as an ascending series, leading up through *Blasia,* by way of *Petalophyllum,* to the undoubted leafy axis of *Fossombronia* (Fig. 1). Nowadays few morphologists read the series in this direction (cf. Mehra[286]), but however one interprets it, there is no doubt that this group of genera in the Metzgeriales illustrates well the lack of any sharp line of demarcation between thalloid and leafy states. A parallel situation is encountered in the Anthocerotales where the epiphytic *Dendroceros* consists of a stout axis and lobed wings of tissue one cell thick. Again, 'quasi-leafiness' may best describe the condition, for surely it is true that in none of these examples except *Fossombronia* do we meet an axis bearing leaves of clearly defined form as we do in the Jungermanniales. We seem rather to be witnessing thalloid form being shaped, modified, extended in different ways by the hand of evolution, although some would argue that these various 'intermediate states' are displaying the last relics of a leafy ancestry.

We must now explore the diversity of form within the two larger orders, Metzgeriales and Marchantiales, in a little more detail. We may begin with the Metzgeriales where the allies of *Pellia* reveal great diversity of gametophyte structure when examined on a world scale. There are forms which consist of a simple undifferentiated thallus, as do all the simpler species of *Riccardia.* There are species, like *R. multifida* (Fig. 1A) in Britain, where the thallus is remarkable for its pinnate branching. *Metzgeria* (of which *M. furcata* is a common bark-dwelling liverwort in Britain) has a thallus with a well-defined midrib and wings (Fig. 1F). *Pallavicinia* is notable for the conducting tissue in its central strand. Although without specialised thickenings, these cells are like tracheids in shape (cf. p. 126). But like *Pellia,* all are thalloid liverworts in which the thallus itself shows no particular differentiation beyond a midrib and, of course, a crop of rhizoids on its ventral surface.

Three different directions of further development (or specialisation) may be noted briefly. The first, a 'trend towards leafiness', need not detain us, for it has been considered above. Suffice it to add a little more detail regarding the two genera *Petalophyllum* and *Fossombronia.* In *Petalophyllum* the leaf-like appendages are two rows of nearly erect lamellae on the upper surface of the thallus. *Petalophyllum ralfsii* (Fig. 1H) is an uncommon plant of wet hollows in sand-dunes. It is also minute, but is worth searching for in order that its unique structure may be seen. True leaves, borne in two ranks on a cylindrical axis, are seen in *Fossombronia,* with several British species. The axis of the Southern Hemisphere genus *Treubia* bears four ranks of appendages, two large and leaf-like, two that

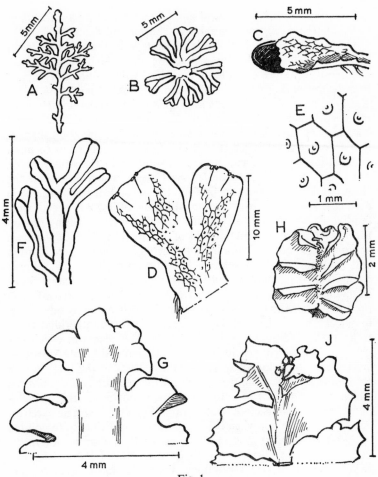

Fig. 1

Form of gametophyte in some Marchantiales and Metzgeriales. A. *Riccardia multifida*—narrow, pinnately branched thallus, no midrib. B. *Riccia sorocarpa*—rosette of dichotomising thallus branches, each with deep median groove. C. *Targionia hypophylla*—on left the two purple-black valves of the perianth protruding beyond apex ventrally (side view of thallus branch). D. *Conocephalum conicum*, view of upper surface of terminal part of branched thallus, the apex in each branch having recently divided again. E. Enlargement of the surface 'areolae' and pores (which are seen faintly in D). F. *Metzgeria furcata*. Ribbon-shaped thallus with clearly defined midrib. G. *Blasia pusilla*, surface view of extremity, showing 'leaf-like' lobes of thallus. H. *Petalophyllum ralfsii*, small plant in which seven of the ± erect, ledge-like lamellae can be seen. J. *Fossombronia angulosa*, extremity of gametophyte branch (with some leaves not yet expanded); differentiation into axis and leaves complete.

are mere scales. It is a genus of great interest, and we are fortunate in having the recent monograph by Schuster and Scott.[388]

The second 'trend' is towards a plant body that is much more highly differentiated, having a creeping axis which bears rhizoids and sends up erect cylindrical branches which ultimately expand into dichotomising fronds. Conducting strands extend throughout. A very good example of this kind of specialisation is seen in *Hymenophytum* although it also occurs in a number of other genera. Experience with *H. flabellatum* (from New Zealand) in cultivation suggests that frond renewal is an annual event. Such forms seem to mark culminating points of gametophyte evolution in the Metzgeriales. They were certainly so regarded by Cavers, who emphasised the close parallels that were to be found in several distinct genera, in each case leading up to a form in which the ultimate branches were fan-like fronds. In *Hymenophytum flabellatum* the resulting growth form, which is a little reminiscent of a filmy fern (*Hymerophyllum*), sporophyte, is adapted to a habitat where high annual precipitation and permanently high relative humidity prevail. In the altered conditions of greenhouse cultivation at Reading the plants failed to maintain the striking appearance they had on arrival, and produced branches that were no longer frond-like, but merely dichotomising. The small, postical cushions that form the antheridial receptacles are an additional specialised feature. The genera in which this 'trend' is best seen are exclusively Australasian and South American.

The third direction of specialisation consists of the total loss of chlorophyll. This condition is found in *Cryptothallus mirabilis* which is thus a plant of unique interest, although in other gametophyte characters it is close to *Riccardia*. Williams[465] gave an excellent account of this plant when it was first found in Britain. He pointed out that the thallus was more succulent than that of *Riccardia*, a fact no doubt correlated with its mycorrhizal habit. The first British record of this saprophytic liverwort was from a locality near Helensburgh, Dumbartonshire, in 1948, but it has since been recorded from many parts of the country. Dickson[103] has summarised its status up to 1968 and also indicated the position in parts of Continental Europe.[104] He sees it as under-recorded rather than rare, for it has to be sought *beneath* the surface litter under birch (*Betula*) or *Sphagnum* on the fringes of bogs. Unique in the biology of its gametophyte, *Cryptothallus* also shows several unusual features in its sporophyte (cf. p. 52). It had first come to the attention of bryologists about 1914, after the buried thallus had been found in a wood near Vienna by Zederbauer who mistook it for prothalli of a species of *Lycopodium*.

Müller[300] recognises six families of Metzgeriales in Europe and all have at least one representative in Britain. The presence or absence of a midrib and of leaf-like appendages furnish two important characters on which family distinctions are based. Another is provided by the diverse position of the sex organs. For although these are on the upper surface of the thallus in *Pellia* and some other genera, they occur on specialised short branches in *Metzgeria* and *Riccardia*. In the former these branches arise ventrally; in the latter they are typically side branches. When one takes into account a number of other characters, including several of importance derived from the sporophyte generation, it is clear that the Metzgeriales comprise a number of small groups, each somewhat isolated from the others. The simpler species of *Riccardia* (such as *R. pinguis* and *R. sinuata* in Britain) might seem to form the logical starting point of the series (but cf. the views of Schuster). Only the palaeo-botanist (cf. Chapter 11) can tell us the age of their simple thalloid form.

As we have seen, *Preissia quadrata*, typifying the more complex Marchantiales, presents the most elaborate thalloid gametophyte to be found among liverworts. Thus it possesses: (1) photosynthetic air chambers; (2) each chamber opening by a barrel-shaped pore with an adjustable aperture (Walker and Pennington[437]); (3) two kinds of rhizoids, smooth and tuberculate, and ventral scales with appendages; (4) antheridia borne within a stalked receptacle; (5) archegonia (and hence sporogonia) borne on a stalked receptacle—the whole mushroom-shaped structure being known when fruiting as the carpocephalum. Each of these five features will be examined in turn to see how it has become modified, or appears simplified, in related members of the Marchantiales.

Probably the large genus *Marchantia* alone, with its bigger (and well-known) male and female mushroom-shaped or umbrella-like receptacles, comes close to *Preissia* on all these five points. In the arrangement of Müller[300] only *Marchantia*, *Bucegia* and *Dumortiera* are placed with *Preissia* in the Marchantiaceae, as represented in Europe. The abundantly gemmiferous *Lunularia*, *Conocephalum*—remarkable for its robust, glossy dark green thallus—and the delicate non-British genus *Exormotheca* are made each the basis of a separate family. In some of these genera, and even more in the simpler Marchantiales such as *Targionia*, *Corsinia* and *Riccia* (again separate families), we can see one or more of the above five features modified or simplified.

(1) *Photosynthetic air chambers.* The classical illustration of reduction in these chambers is furnished by *Dumortiera*, in which the response to differing environmental conditions was investigated

long ago by Coker.[89] Excessive moisture induces atrophy of the
chamber as such, the roof being lost entirely whilst only scattered
short chains of green cells spring from where the floor has been.
In well-watered *Dumortiera hirsuta* grown in shade I have found
indeed that the chambers are habitually lacking, even in the youngest
parts of the plant, an epidermal layer of photosynthetic cells covering
the upper surface of the thallus. In *Preissia* the photosynthetic
region occupies only about a quarter of the midrib thickness of the

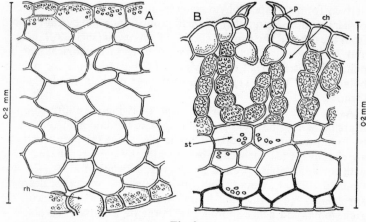

Fig. 2

Comparison between thalli of *Pellia* and *Preissia* in transverse section.
A. *Pellia fabbroniana*, T.S. midrib region. rh. origin of rhizoid. Note that
chloroplasts are concentrated mainly in upper and lower epidermis.
B. *Preissia quadrata*, T.S. wing region, showing barrel-shaped pore (p.)
air chamber (ch.) with columns of chlorophyllose cells. Below some cells
store starch (st.). The walls of cells on ventral surface (shown dark) hold
anthocyanin pigment (deep purple in life).

thallus; and one line of departure is found in those species where
more than half the thickness is occupied by an irregular network of
chlorophyllous tissue. This produces a complex chamber system
in place of the single layer of chambers seen in *Preissia* and *Marchantia,* but the walls of the chambers themselves provide the green
tissue and there are here no additional plates or filaments. This is
seen in *Reboulia,* and here and there elsewhere. In the majority of
species of *Riccia* the whole upper surface of the thallus consists of
close-set columns of green cells. A pear-shaped colourless cell
commonly terminates each column and the whole system is freely
provided with fine air channels. There are no chambers and pores
as such here. The thallus segments that compose the neat rosettes

of *Riccia* species are narrow and often U-shaped or V-shaped in transverse section; the delicate columns of photosynthetic tissue can be given some protection in dry conditions by the in-folding of the thallus margins.

(2) *Type of pore*. Only in *Preissia* has the barrel-shaped pore been shown to be adjustable in aperture size; but rather similar pores are found in *Marchantia* and several other genera. In *Reboulia*, which belongs to a family where the air pores are normally simple, barrel-shaped pores of remarkable size (seven to eight cells in depth) occur in the photosynthetic region of the carpocephalum. The same is true of *Conocephalum*.[73] Simple pores prevail in many of the lower Marchantiales—*Targionia, Corsinia* and others. Even where they are simple, however, the pores vary in detailed structure and hence in the appearance presented in surface view. Thus the three calcicole genera *Clevea, Sauteria* and *Peltolepis* which represent the family Cleveaceae in Europe, were formerly known as the Astroporae ('star-pored'). The pores owe their star-like form to the thick walls of the cells which surround each as a single well-defined ring. In *Reboulia* and its allies each simple pore is bounded by several concentric rings of cells. The pore becomes a very unspecialised structure in *Corsinia* and in those species of *Riccia* (such as *R. crystallina* and *R. huebeneriana*) where it persists.

(3) *Ventral scales* are most prominent, and functionally most important, close behind the apex of a thallus segment. Cavers[73] figured prominent scales lying over and effectively protecting the apex in *Conocephalum*. They are seen even better in species of *Plagiochasma, Grimaldia* and *Targionia,* plants of dry habitats in which the scales together with the infolding thallus margins are clearly of protective value. Ventral scales are delicate plates of tissue only one cell thick. They are closely overlapping, and characteristically set in two rows. The scales are often deep purple. In most of the higher Marchantiales they are appendaged, although the appendage is sometimes hard to demonstrate. Appendages are also found in *Targionia,* but have been lost (or perhaps they never existed) in *Corsinia* and *Riccia*. Here too the scales themselves are either scattered or in a single irregular row. At times, especially in aquatics, they may be difficult to see at all. Two kinds of rhizoids are a distinguishing mark of the Marchantiales as a whole.

(4) *Antheridiophore*. In this feature a clear-cut series can be seen which may however be interpreted as running in either direction. *Preissia* and *Marchantia* have a stalked, green or purplish structure like a miniature mushroom. In *Conocephalum,* as in *Reboulia* (Fig. 18E) and most of its allies, the antheridia are grouped in

sessile, cushion-like receptacles. In *Corsinia* they are irregularly grouped along a portion of the thallus midrib; whilst *Riccia* produces antheridia singly at various points on the surface of the thallus in or near the median line. Often the site of an embedded antheridium is marked by a projecting turret of surrounding thallus tissue.

(5) *Archegoniophore*. Stalked archegonial receptacles (and thence carpocephala, or fruiting heads) are much more widespread than the corresponding stalked antheridial structures. They are found throughout the Marchantiaceae and all closely related families; but the number of archegonial groups in the rayed head varies from eight in *Marchantia polymorpha*, through four in *Preissia* and *Lunularia*, to two in the diminutive *Exormotheca*. They are found also in *Reboulia* and its allies; again in *Clevea, Sauteria* and *Pelto-lepis*. These two groups run parallel in some respects; thus *Plagio-chasma* (related to *Reboulia*) and *Clevea* are both unusual in that the archegoniophore does not remove the growing power of the apex and hence appears to spring from a point well back on the upper surface of the thallus. *Targionia* is unusual in that the archegonia arise ventrally behind the thallus apex, and the sporophyte is enclosed by a pair of close-fitting dark purple structures like the valves of a mussel shell (Fig. 1c). In *Corsinia* the archegonia are borne on specialised parts of the upper surface of the thallus; associated with them is a flap of tissue which may be either an incipient or a vestigial archegoniophore. Finally, in *Riccia* the archegonia like the antheridia are scattered, solitary and embedded in the thallus.

From early times (Leitgeb)[258] much emphasis has been laid upon the detailed structure of the carpocephalum in the different genera of the higher Marchantiales in an attempt to arrive at a satisfactory classification of them. Thus, the stalk may have no furrow (as in *Clevea* and *Plagiochasma*), bear a single furrow with rhizoids as in *Reboulia, Conocephalum* and many others, or bear two such furrows, as it does in *Preissia, Marchantia* and a few other genera. Also, the archegonia may occur singly in each involucre as they do in most Marchantiales, or they may be clustered, as in a few specialised genera including *Marchantia* itself. A specially interesting situation is seen in the rather recently discovered genus *Neohodgsonia* (Persson, 1954[331]) for in its form and the manner of its branching, the 'head' of the carpocephalum reveals its essentially thalloid nature more clearly than in any other genus.

This completes our survey of thallus structure and insertion of gametangia in the Marchantiales. This topic is one worth entering in some detail, for not only is the complex thallus of some members of this group without parallel elsewhere among bryophytes, but furthermore these liverworts are exceptionally easy to grow and, once a greenhouse collection has been started, they are freely available for study. Any evolutionary interpretation of the facts will be deferred until the sporophyte generation has been reviewed (cf. Chapter 4).

This chapter may conclude with accounts of two remarkable and puzzling genera of thalloid liverworts—*Monoclea* and *Monocarpus**. It will be convenient to refer briefly to both gametophyte and sporophyte generations. *Monoclea forsteri* (Fig. 3D,E) was brought back from New Zealand by Forster on one of Captain Cook's voyages and erroneously named *Anthoceros univalvis*. W. J. Hooker in 1820 made it the type of a monotypic genus, *Monoclea,* naming it after its discoverer. In 1858 it was the subject of a detailed investigation by Gottsche and much later it received attention from the American morphologist, D. H. Campbell,[63] who was the first to place it in the *Marchantiales*. Johnson[219] and Cavers[72] also studied and commented upon the genus, the former making use of material of a second species, *M. gottschei,* from Jamaica. On balance the relationship with Marchantiales was endorsed, only to be challenged at a later date by Schiffner.[377] Evans, moreover, was inclined to the view that the genus was nearer to the Metzgeriales and some others have shared this view. In 1961 Proskauer[344] reviewed the position, and whilst rejecting as inadmissible much of the earlier evidence, concluded that *Monoclea* belonged to the Marchantiales. Finally, Schuster has assigned it to an order of its own.

Features of *Monoclea* which support a relationship with *Pellia* are: (1) absence of any trace of ventral scales; (2) no air-chamber system; (3) insertion of the archegonia on a sloping surface of thallus where they are protected by a collar-like involucre; and (4) the most striking character (and the only one which Proskauer admits to be of any weight), a seta several centimetres long. Then there are the following six characters which have been cited at various times in support of its relationship with the Marchantiales: (1) possession of both smooth and tuberculate rhizoids, which differ also in direction of growth; (2) presence of oil body cells and mycorrhizal cells within the massive thallus. These two characters alone are admitted by Proskauer. Four others, of more doubtful value, are: (3) male receptacle cushion-like, as in many genera of Marchantiales; (4) six rows of cells in the archegonial neck; (5)

---

* The generic name *Monocarpus* was rejected by Proskauer[347], who renamed it *Carrpos*; but Schuster[384] has retained *Monocarpus* which he makes the sole genus of the family Monocarpaceae.

'Marchantioid' embryology, with an octant stage represented; and (6) a single layer of cells in the capsule wall. The fifth point is to some extent discounted because many genera of Marchantiales do not themselves pass through an octant (ball of eight cells) stage in their embryology. Schuster[384] has recently emphasised the importance of the 'compact cell body' formed on the germination of the spore as something quite distinct from what obtains in the Marchantiales. He also alludes to the complex structure of the capsule wall. These features, taken with other considerations, have moved him to accord it full ordinal status.

Summarising, one may say that *Monoclea* combines a certain superficial resemblance to *Pellia* with many fundamental points of departure from the structural pattern of the Metzgeriales. Nor does it fit well into the Marchantiales. Proskauer believed that Burgeff's massive researches[60] on the mutation potentialities of *Marchantia* were relevant to the problem. He wrote:[344] 'Prior to Burgeff it would have been too daring to suggest a relationship between *Dumortiera* and *Monoclea*.' Such a suggestion depends upon the assumption that *Monoclea* is a reduced descendant of carpocephalous ancestors and has struck some other bryologists as far-fetched, especially when one invokes as the ancestor a genus such as *Dumortiera* which itself bears many marks of reduction. On balance, the erection of a new order, Monocleales, may be the best solution. One conclusion at least seems inescapable; that the genus *Monoclea*, with its two species showing markedly disjunct distribution, and its odd combination of characters, must be a very ancient type of thalloid liverwort. As Parihar[320] reminds us, it is also exceedingly large, with a thallus attaining 20 cm long and 5 cm wide.

Our second example of an anomalous thalloid liverwort (posing a problem as to where it should be placed systematically) is a very different one. For it is a minute plant with a thallus diameter normally of less than 2 mm; and it represents a recent discovery. These tiny plants were found by D. J. Carr in 1955 on the bare mud of a saltpan in north-west Victoria, Australia. He named the new genus *Monocarpus*, from the single sporophyte borne by each fertile specimen. Carr[68] gave an excellent account of it and, discussing its relationships, saw some resemblances to *Sphaerocarpos*, others to certain genera of the Marchantiales. His figure of a vertical section of the whole fertile plant is reproduced here (Fig. 3A–B) and the student may ponder for himself the questions which it raises. Antheridia were discovered later by Proskauer when he was growing *Monocarpus* from spores. They are broadly ovoid and long-stalked. The species is monoecious.

The sporophyte (cf. Chapter 4) resembles that of *Sphaerocarpos* but is also akin to those of some Marchantiales (e.g. *Corsinia*). The vegetative part of the gametophyte appears to be extremely simple and certainly recalls *Sphaerocarpos*; and even the elaborately lobed structures obtained by Proskauer[348] when he germinated spores and cultivated the plant at a later date are not out of line with that genus. The globose receptacles, however, with their chambers, and the pores composed of two tiers of cells, are strong pointers to

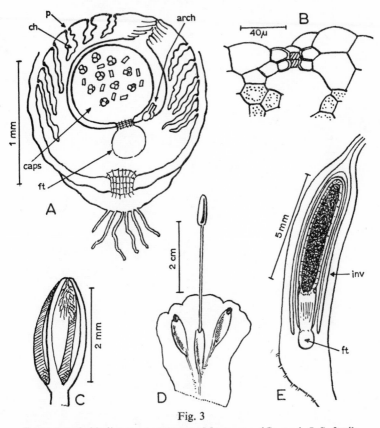

Fig. 3

Some remarkable liverwort genera. A. *Monocarpus* (*Carrpos*). L.S. fertile plant, somewhat diagrammatic. Note simple thallus, bearing rhizoids, and stalked, globose, fertile structure, with chambers and pores. B. Section through 2-tiered pore (which leads into air chamber). C. *Cryptothallus*, dehisced capsule. D. and E. *Monoclea*. D. Part of thallus, with sporophyte, showing long seta and capsule. E. L.S. sporophyte, younger stage. arch. old, unfertilised archegonia; caps. capsule; ch. air chamber; ft. foot; inv. involucre; p. pore. (A. and B. after Carr; C. after Williams; D. after Cavers; E. after Proskauer) the Marchantiales. The solution found by Schuster[384] has been to make *Monocarpus* (*Carrpos* Prosk.[347]) sole genus of a distinct family, Monocarpaceae, of the order Marchantiales. Of considerable interest is the discovery, in 1968, of *Monocarpus* in South Africa (Schelpe)[375].

3

GAMETOPHYTE OF LEAFY LIVERWORTS

The leafy liverworts (Jungermanniales of Schuster, Jungermanniales Acrogynae of many earlier systems) are by far the richest in species of the orders of liverworts. Müller has estimated that 84% of all liverworts are leafy forms, and nearly all are members of this order. A fundamental feature of the gametophyte here is its growth from an apical cell which takes the form of an inverted pyramid consisting of a base and three sides from which derivatives are cut off. With the exception of a few genera (e.g. *Pleurozia*) this so-called three-sided apical cell occurs universally in the group, and since the derivatives from each cutting face contribute to both stem and leaves, the fundamental leaf arrangement is three-ranked. Contemporary morphologists regard a radially symmetrical leafy shoot, formed in this way, as primitive in the order and see all dorsi-ventral shoot systems as derived from it. Thus, we find in the check-list of Jones[221] and in the full treatments of Arnell[16] (for Scandinavia) and Müller[300] (for Europe) that those families and genera with a nearly perfect radial symmetry of the leafy shoot are placed at the beginning, and they are followed by others which are considered to have been modified to varying extents.

Nearly perfect radial symmetry of the leafy shoot is quite rare in the Jungermanniales and only a few British genera show this feature. Among them, two good examples are *Herberta* and *Anthelia*. *Herberta adunca** (Fig. 4A) is a prominent hepatic on mountain ledges in some parts of western Britain, conspicuous on account of its orange-brown colour and robust, erect habit. It may easily be

* *Herberta adunca* (Dicks.) Gray = *H. hutchinsiae* (Gottsche) Evans of earlier editions.

mistaken for a moss at first glance, but closer inspection reveals the three ranks of leaves, all much alike in size and each cleft to the base into two long tapering lobes. *Anthelia julacea,* a plant of mountain springs and stream beds, and *A. juratzkana,* a rather rare species in Britain, characteristic of late snow areas on high mountains, are much smaller plants, but the minute leaves show the same nearly symmetrically three-ranked arrangement and each is again deeply cleft into two acute lobes. This type of leaf lobing finds its explana-

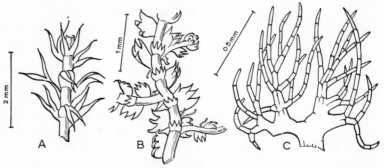

Fig. 4

Gametophyte of Jungermanniales. A. Part of shoot of *Herberta adunca,* showing three ranks of leaves of almost equal size. B. Part of shoot of *Lepidozia reptans,* leaves in three ranks but mid-ventral rank (under-leaves) of noticeably smaller size. C. Leaf of *Trichocolea tomentella,* the two lobes further subdivided and bearing numerous filamentous projections.

tion in another characteristic attribute of the Jungermanniales, that each leaf, from a very early stage, grows by means of two distinct apical growing points. This readily leads to a bi-lobed or cleft leaf at maturity; by contrast a simple, entire leaf of rounded form is relatively uncommon in the group as a whole.

The row of leaves lying in the mid-ventral line is given a special term—amphigastria or 'underleaves', and even in *Anthelia* these leaves are slightly smaller than their lateral counterparts. In such genera as *Trichocolea* and *Lepidozia* there is a greater difference between the sizes of amphigastria and lateral leaves, and in these genera the habit is prostrate or ascending rather than erect. These liverworts, and their immediate allies, illustrate another develop-ment, namely the tendency for the leaf to be divided into three or four lobes. The three or four finger-like lobes can be seen well, under a lens, in the minute leaves of the richly branched cushions of *Lepidozia reptans* (Fig. 4B), a common plant of acid woodland banks. In *Ptilidium* each leaf lobe bears numerous filamentous lateral appendages, and this tendency reaches its ultimate expression in such a plant as *Trichocolea tomentella*. Here the finer ramifications of the leaf lobes (Fig. 4C) remind one of a richly branched filamentous alga. This distinctive, pale green leafy liverwort is found in fen carr and certain other habitats where a permanently high relative humidity can be relied upon.

The adoption of a horizontal or plagiotropous shoot system is very widespread among the Jungermanniales, but it is interesting that this appears to have been achieved in two ways. These give rise to the so-called succubous and incubous leaf arrangements. Close to the apex the leaf rudiments are almost transversely inserted or, as Buch[56] has pointed out, they may be already oblique at this early stage, becoming more so as they mature. In the larger group of leafy liverworts the lower, or ventral, edge of the leaf is carried forward relative to the upper or dorsal edge and the leaves on this resulting shoot will then overlap one another in a succubous manner. When the opposite condition prevails and the upper, or dorsal, edge of the leaf is carried forward the resulting overlap is of the incubous type. Fewer genera exemplify this condition. In both instances we are witnessing a precise growth adaptation to the prostrate or ascending habit, and one which results in the leaf surfaces being brought into a good position to receive the incident light. It must not be forgotten that there will have to be a compensatory growth at some stage in an opposing sector of stem tissue; otherwise the stem itself would become sharply curved, which in fact is not ordinarily so. As an example of the succubous condition one may take *Lophocolea* (Fig. 5B); of the incubous condition, *Bazzania* (Fig. 5A). Both genera possess amphigastria only a small fraction of the size of the lateral leaves.

Lophocolea and *Bazzania* are both large genera, and both are distributed throughout the world. *Lophocolea heterophylla,* with decaying logs as its principal habitat, is easily obtained and will serve to illustrate many features that are typical of the order as a whole. The delicate, little-differentiated stem, the two-lobed leaves (here intermixed with others that are nearly entire), the groups of colourless, unicellular rhizoids and the deeply cleft amphigastria are all features that are readily seen. Each leaf is a single layer of large, nearly isodiametric cells. Archegonia arise apically, a position characteristic of the order as a whole (hence the name Acrogynae), the apical cell being involved in the process. Antheridia occur in the axils of specialised concave leaves just behind the apex and the plants are usually very fertile. A protective envelope (the perianth) surrounds the archegonia and later the developing sporophyte. It is a highly characteristic feature of the order and will be discussed later.

Bazzania is best exemplified in Britain by a robust species, *B. trilobata,* which is often abundant in mountain woods in the west and north. The arched, freely branched leafy shoots form conspicuous mats or cushions; and each shoot bears a certain number of nearly bare, descending branches termed flagella. The incubously

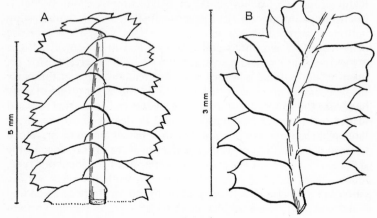

Fig. 5

The extremities of two contrasted types of leafy shoots viewed from above (i.e. antical view, with underleaves concealed beneath stem). A. *Bazzania trilobata*, leaves incubous, each leaf trifid; B. *Lophocolea heterophylla*, with succubous insertion of leaves, most of which are deeply bifid.

arranged leaves have three little teeth at their blunt tips; the small amphigastria are multifid. This species is dioecious and only rarely fertile. These two contrasting forms may be used as the foundation on which to build one's knowledge and experience of the host of liverwort genera that like them have adopted a plagiotropous mode of growth with obliquely inserted leaves. We must now consider one or two other variations of leaf arrangement or leaf form that have arisen within the order.

The logical outcome of reduction in size of amphigastria (if indeed reduction has occurred from a once radially symmetrical ancestral form) is the total elimination of these underleaves. A common British liverwort offering a good example of this is *Lophozia ventricosa*. Many Lophoziaceae show amphigastria which have the appearance of being vestigial, for they are minute and confined to positions just behind the growing apex. In the two closely allied mountain genera, *Marsupella* and *Gymnomitrion*, there is no trace of underleaves, although the habit of the commonest British species of *Marsupella, M. emarginata,* is often nearly erect. *Bazzania* was noted as unusual in its trifid leaves. Leaves in some genera are simple and rounded in outline, the margin entire or nearly so. Thus, *Nardia scalaris, Mylia taylori* and *Odontoschisma sphagni* have simple leaves with entire margins; *Plagiochila asplenioides* has simple, markedly decurrent leaves with minutely toothed margins. Generic features here are the recurved antical margin and arched postical insertion of the leaf.

It remains to consider the most striking development that occurs in the bilobed leaf, with its two independent growing points. There are two important lines along which such further development has taken place, leading to leaves of more complex form than those in the examples cited so far. First, there are plants in which the leaves have become folded, to produce a keel-like form in the basal part of the leaf, whilst distally the smaller antical lobe lies across the large postical lobe. This is seen in *Diplophyllum* and in many species of *Scapania.* Neither genus has any trace of amphigastria. In some species of *Scapania* there is little difference in size between the two lobes. There is a gradual transition to the rather striking shoot pattern of *Diplophyllum* (literally, double leaf)—Fig. 6A.

An even more striking modification occurs when the two growing points of the young leaf remain distinct from the outset, with the

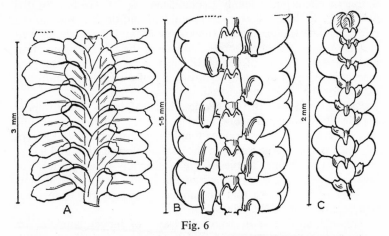

Fig. 6

Further examples of structural diversity in the leafy shoot (gametophyte) of Jungermanniales. These three illustrate the 'complicate-bilobed' leaf. A. *Diplophyllum albicans,* part of shoot viewed from above and showing smaller size of antical leaf lobes. B. *Frullania tamarisci,* viewed from below; postical lobes smaller and helmet-shaped, underleaves present. C. *Lejeunea cavifolia,* viewed from below; postical lobes small, saccate, and underleaves again present.

result that they develop into two structures almost free from one another and entirely different in form. A good illustration of this is provided by *Frullania,* which is not only one of the largest genera of leafy liverworts, with perhaps 500 species, but is easily obtained in Britain. *Frullania tamarisci* is abundant on rocks and trees in north and west Britain; *F. dilatata* is a generally widespread epiphyte. Either will illustrate the remarkable leaf structure, with the larger, antical lobe expanded and the smaller, postical lobe taking the form of a helmet-shaped pitcher on a short stalk (Fig. 6B). A minute projection, the stylus, grows out from the stalk. It is instructive to examine younger parts of the shoot in *Frullania,* for then stages in the development of the pitchers may be seen. The postical lobe may then appear concave and shell-like in form, but far removed from the perfect helmet of maturity. In *Frullania* there is also a row of amphigastria. The ecological significance of these pitchers has been discussed fully by Goebel[148] and by others more recently. Parihar is probably right in suggesting that their main function is to catch water that would be trickling down a tree-trunk or rock face after a shower. In tropical species of *Colura* and *Pleurozia* the postical lobe of each leaf takes the form of a more perfect trap and Müller alludes to its use in trapping minute animals after the fashion of the bladderwort (*Utricularia*). Many members of the Southern Hemisphere family Lepidolaenaceae (cf. Grolle, 1967[155]) exhibit single 'helmets' on each lateral leaf and double ones on the bilobed underleaves. Thus, *Gackstroemia* and *Lepidolaena* can surpass *Frullania* in the elaborate character of their shoot organisation and represent a peak of complexity in this respect. Such trapping mechanisms carry at least the possibility of an enhanced nitrogen supply for the plants in question.

Porella, with five British species, has leaf lobes which appear practically distinct from one another, but the small postical lobe is not saccate. In the enormous family Lejeuneaceae (many genera of which play so large a part in the colonisation of bark and leaves of trees in the humid tropics) there is normally a saccate form of postical lobe. This, often known as the lobule, has a critical morphology of its own so that taxonomically it is a very significant structure (cf. Jones,[222] Greig-Smith[153]).

Although so diverse in detail, the leaves of Jungermanniales are in some ways rather constant and limited morphologically. With few exceptions, each leaf is composed of a single layer of cells and there is no kind of midrib differentiation. *Pachyglossa* (cf. Herzog and Grolle[183]) with leaves two to six cells thick and *Chondrophyllum,* described by Herzog[182] in 1952, are two out of a small handful of Southern Hemisphere genera where the whole leaf is multistratose.

The nearest approach to a midrib is a line of specialised cells running up the middle of the leaf. We see one instance in the 'vitta' of *Diplophyllum* and *Herberta*. A very different one is illustrated by the line of enlarged cells (ocelli) found in a few species of *Frullania*. Otherwise, cells tend to be of a fairly constant type, most often rounded-hexagonal and nearly isodiametric, throughout the leaf.

Rhizoids, too, are subject to little structural variation. They are simple, tubular outgrowths from specialised cells. According to Müller,[300] multicellular rhizoids are known only in *Plagiochila paradoxa* and the genus *Schistochila*; but to these Fulford[134] adds *Vetaforma* (first described in 1960). Rhizoids are sparse in some erect-growing leafy liverworts. In many genera (e.g. *Lophozia*) they can occur along a continuous length of the ventral surface of the stem; but elsewhere (e.g. *Lophocolea*) rhizoid-bearing cells are localised—frequently at positions near the insertion of the under-leaves. Schuster[384] points out how, in Lejeuneaceae, Frullaniaceae and Porellaceae, a group of cells actually on the base of the underleaf remains meristematic after surrounding cells have matured, and thus comes to be rhizoid-bearing. The groups of rhizoids which result in these instances affix the delicate plants tenaciously to the substratum. The situation in *Radula,* where rhizoids spring from the centre of the postical lobe of the leaf, is quite exceptional. Paraphyllia (small appendages of varied form) are rarely found; but they clothe the stem of *Trichocolea tomentella* where they probably increase the facilities for capillary water movement. There is room for further research on the question of how far a particular pattern of rhizoid distribution is genetically determined and how far it fluctuates with changing environment. Bark-dwelling liverworts will always demand a firm attachment.

The structure of bryophyte sex organs will be taken up in Chapter 8, but we may now turn to examine that characteristic protective envelope, the perianth. It belongs to the gametophyte generation, although functionally it is associated with the sporophyte. An erect, tubular or funnel-like structure, it is often the most conspicuous feature of small liverworts, such as *Cephalozia* species, when these are fertile. Morphologically, it is sometimes hard to interpret, but taxonomically it has proved very useful.

It is often possible to harmonise the form of the perianth with the idea that it consists of either two or three fully united, modified leaves. Cavers[74] summarised well the four chief arrangements that can result, on this interpretation, and the accompanying diagram (Fig. 7) will make this clear. The four possibilities, with examples, are: (1) three leaves participating, resulting perianth three-angled along lines of union of leaves, e.g. *Lophocolea*; (2) the same, but each leaf folded, resulting perianth three-angled along the median lines or keels of component leaves, e.g. *Frullania*; (3) two leaves participating, not keeled and hence forming a laterally compressed perianth, e.g. *Plagiochila*; (4) the same, but the leaves folded or keeled, hence giving a dorsiventrally compressed perianth, e.g. *Scapania*. Since wherever three leaves participate one is in the mid-ventral line (corresponding with amphigastria) it follows that in type (1) there will be an angle in the median dorsal (or antical) position; in type (2) there will be an angle in the median ventral (or postical) position.

The above is a short, simplified statement that will enable the student to grasp the essentials. For a fuller treatment he should consult Schuster[384] where (pp. 548–57) a lavishly illustrated account is given. We can consider only a few additional points. Firstly, the younger stages in perianth development often reveal quite clearly its origin from three 'leaves' even when this is obscured at maturity. Thus, the mature perianth in such genera as *Lophozia*, *Solenostoma* and to some extent *Cephalozia* is a tubular structure variously pleated (plicate) distally but not obviously made from a fixed num-

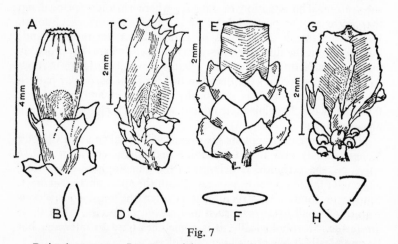

Fig. 7

Perianth structure. Jungermanniales. A. Tubular perianth of *Lophozia ventricosa*. B. Plan of laterally compressed, two-member perianth of *Plagiochila*. C. *Lophocolea heterophylla*. D. Plan of this three-member perianth. E. *Scapania undulata*. F. Plan of this dorso-ventrally compressed, two-member perianth. G. *Frullania dilatata*. H. Plan of this three-member perianth, with folded members and keel in mid-ventral line. In A, C, E and G an indication is seen of the 'bracts', of modified form, which stand in lateral positions just below the perianth, and in C and G the 'bracteoles' —larger than, and different from, ordinary underleaves.

ber of components. Again, within Lejeuneaceae, although in some genera (e.g. *Mastigolejeunea*) the '*Frullania*' structure obtains, most others are variously altered, often through the smaller mid-ventral component remaining unfolded and the appearance of accessory folds in new positions. Amid all its diversity, the perianth stands as an obviously important protective structure and one that has been much used in taxonomy, at both family and generic levels. It may be mentioned, finally, that at the time of fertilisation of archegonia it is frequently only beginning to develop. Schuster[384] reminds us that at this stage the terminal fringe of 'cilia' (that is often present) may play the vital role of moisture conservation around the archegonia when the greater part of the perianth is still at the formative stage. In a few genera, e.g. *Gymnomitrion,* a perianth is lacking. Modifications of perianth structure associated with marsupial development are taken up in Chapter 4.

On female shoots, and situated between vegetative leaves and perianth, are the so-called 'bracts' and (sometimes) 'bracteole'. These correspond respectively with lateral leaves and amphigastria. They tend to be larger than vegetative leaves, but in certain cases, e.g. *Chiloscyphus* and *Calypogeia,* they are smaller. Like the perianth, they tend to undergo modification where a marsupium is formed. In *Marsupella* and *Plectocolea,* there is partial fusion between bracts and perianth. *Frullania* displays both bracts and bracteole of distinctive form for particular species. Male bracts commonly have a specialised form, being markedly concave where they surround the antheridia. The appearance is most striking when a whole branch is male, for then it will often have a distinctive, catkin-like form.

Branching in Jungermanniales is never a true dichotomy (forking) brought about by the symmetrical cleavage of an apical cell, although we find apparent dichotomies, e.g. in *Bazzania trilobata.* The systems of branching which occur are complex and varied. They have been fully described and figured by Evans.[119] Terminal branching types, of which four were recognised by Evans, are always intimately related to the mode of origin of derivatives from the three cutting faces of the apical cell. By far the commonest is the *Frullania* type in which the branch comes from the whole ventral half of a lateral segment of the apical cell. Hence, immediately above the point where the branch emerges, one finds a leaf which lacks the product of this half-segment,—in *Frullania* itself the helmet-shaped postical lobe. The *Radula* type, which occurs also in many Lejeuneaceae, involves no such modification of an associated leaf, since only a small part of the ventral half of a lateral segment takes part in the branch origin. Evans's third and fourth types are rare. His third (known as the *Microlepidozia* type) involves replacement of the antical (dorsal) half segment by a branch, so that the antical part of the associated leaf is missing. Thus it is the reverse of what happens in *Frullania*. In Evans's fourth type—found in *Acromastigum* and a few other genera—the branch replaces either half of an under-leaf.

All these are 'terminal' (or acroscopic) modes of branching and, although the branch rudiment may long lie dormant (as it commonly does in *Radula*), the branches are exogenous in origin so that they appear smoothly continuous with the 'main axis' on which they are borne. A somewhat transitional situation, however, obtains in the family Lejeuneaceae. Altogether contrasting with the above examples are true intercalary branches. These also are widespread in leafy liverworts, and are distinct in being long delayed and showing no definite sequence. Their commonest position is in the axil of a leaf (lateral leaf or underleaf) and they are endogenous in origin so that a minute 'collar' will tend to appear where the branch has broken through the superficial cortical tissues. This whole subject, which is significant both in determining the habit of a liverwort and in provid-ing clues to relationships, has in recent years been exhaustively reviewed by Schuster[384] and by Crandall*. Because some of the old descriptive terms have been given different meanings by various authors, Crandall prefers to consider the situation afresh, genus by genus. She finds greater diversity than has been hitherto appreciated, but concludes that the range of branching types in any one species appears to be genetically controlled.

Leaf cell dimensions, extent of thickening deposited in the angles of the cells (trigones) and the type of oil body (if any) all provide important taxonomic characters. Among British examples the diameter of the leaf cells varies from 15 μ or less in species of *Cephaloziella* to over 50 μ in *Mylia* and some species of *Cephalozia*. The plate of cells that forms the leaf lacks air-filled intercellular spaces, but large trigones are prominent in many genera, e.g. *Gymnomitrion, Mylia,* etc., and it is uncommon for such thickenings to be absent altogether. Oil bodies of varied form, usually somewhat larger than chloroplasts and quite colourless, occur in a high pro-portion of genera (cf. Inoue[280]). According to Müller[300] they are usually mixtures, either of terpenes and terpene alcohols or of sesquiterpenes and sesquiterpene alcohols. Their form is often characteristic of particular genera, or even species. Rather little is known of their function.

The young stages of the gametophyte are of interest from two

*For a recent, very full study of this subject see Crandall, B. J. (1969), 'Morphology and the development of branches in the leafy Hepaticae', *Beihefte 3, Nova Hedwigia,* 30, p. 261.

points of view, the morphogenetic and the evolutionary. The first of these is considered later (Chapter 9); the second may fit into the present context, for before we conclude this outline of gametophyte structure in Jungermanniales we must refer briefly to the possible phyletic sequences in this very large order. Also, where do they stand in relation to other groups of liverworts? Juvenile stages might throw light on these questions (cf. Fulford[132]), especially if there is a possibility of the plants revealing a hint of ancestral form in the early development of the individual. Fulford (and others) have made it clear that filamentous sporelings, cell plates and various kinds of 'cell mass' are characteristic of different groups (cf. p. 119). Such forms of plant body might be seen as reflecting a remote algal ancestry. Indeed, long ago Goebel[148] referred to what he called 'reversion to thallus form' among Jungermanniales, giving as examples exceptional plants such as *Pteropsiella frondiformis* and *Zoopsis argentea*. In these instances the whole vegetative part of the plant takes the form of a band-like or strap-shaped thallus and leaves are mere vestiges; yet the reproductive shoots are leafy in the normal manner. Contemporary morphologists are inclined to see such cases as illustrative of a downgrade sequence from leafy to thalloid state. The rudimentary leaves are seen as a survival from a formerly more complete state. Another example, comparable with the above, is *Schiffneria*, which Schuster describes as having the 'axis noticeably flattened and leaves reduced to entire lobes co-extensive with the semithallose cauline region'.

It is characteristic of leafy liverworts that between the very young sporeling and the mature shoot system they commonly display a juvenile stage. In some instances, where the mature shoot bears two ranks of leaves and rudimentary underleaves (e.g. *Mylia*, quoted by Schuster[384]), the juvenile stage displays three ranks of leaves of nearly equal size. This clearly suggests that mature gametophytes that display anisophylly have been secondarily modified and it adds one more 'strut' to support the argument that the earliest type of leafy gametophyte was radially organised and bore leaves in three equal ranks. Recent research (Fulford,[134] Schuster[384] [386]) has revealed numerous genera, e.g. *Archeophylla, Archeochaete* and *Herzogiaria* from the Southern Hemisphere, which are monotypic or nearly so and embody a series of features which seem to be clearly associated with the primitive leafy state. Among these are a diversity of branching methods and a tendency for antheridia to appear in the axils of all three ranks of leaves. Others relate to topics which will be taken up in later chapters. The closest British relative of these plants is *Blepharostoma trichophyllum*.

It remains to consider two groups which are 'leafy' but yet do not belong to the order Jungermanniales. Both are highly significant despite their small size. The first is the order Calobryales. In this country we have long known the widely distributed Northern Hemisphere species *Haplomitrium hookeri* as a British rarity, and older authors recognised that a second genus, *Calobryum* occurred in markedly disjunct fashion in many other parts of the world. The two genera have now been united (by Schuster[375]) under *Haplomitrium,* of which eight species are currently recognised (cf. Fig. 26 p. 185). They bear a superficial resemblance to the Jungermanniales—indeed one species lay for forty years 'hidden' in that order, to a genus of which it had been wrongly assigned in the first instance. Because they display radially organised leafy shoots with a high degree of differentiation (e.g. prostrate, erect and descending components) and a central 'conducting strand' some earlier authors (including Smith[403]) regarded them as something of a climax in leafy liverwort evolution. Such a viewpoint, as Schuster has made abundantly clear, is quite unwarranted by the facts. Indeed, he lists an impressive series of undisputably primitive characters for the Calobryales. These involve the radial organisation itself, the mode of branching, 'biological' characters such as a lack of drought resistance and an absence of asexual reproduction; also (very significant) primitive sex organ structure (cf. Chapter 8). There are further decisive characters relating to the sporophyte.

The paradox is that these plants *look* like Jungermanniales, but in fact are almost certainly closer to Metzgeriales. When a third rank of much smaller leaves occurs (as it often does in *Haplomitrium*) they are not amphigastria, but in fact represent a *dorsal* row of appendages. The leaves themselves are of a rather ill-defined form and there is an absence of any indication of growth from two discrete growing points in the young leaf. Again, although archegonia may be aggregated into a terminal head the apical cell seems not to be involved, so that *Haplomitrium* is essentially anacrogynous, not acrogynous, as are all true Jungermanniales. Finally, a notable characteristic lies in the total absence of rhizoids. This feature is linked with the presence of a remarkable series of leafless, positively geotropic branches (Fig. 26, p. 185) which Grubb,[158] in a recent contribution, has seen fit to term 'roots'.

On the first intimation of the completely new genus *Takakia* (Hattori and Inoue, 1958[169]) Schuster suspected that a 'class parallel with Musci, Hepaticae and Anthocerotae' was 'at hand'. Later, in the classification which we are following, he saw it as worthy of ordinal rank within the Hepaticae; but soon afterwards (Schuster, 1966[385]) he was prepared to reduce it to the suborder Takakiinae, within Calobryales. Thus, closer acquaintance with the genus has

caused competent observers to regard its peculiarities as less fundamental in character than was at first suspected. Many features,
indeed, it shares with *Haplomitrium,* among them radially organised
shoots (Fig. 26), total lack of rhizoids, axillary slime papillae and a
massive, undoubtedly primitive type of archegonial structure.

The obvious unique feature of *Takakia* lies in the character of the
appendages which the axis bears (Fig. 26c–f, p. 185). Earlier accounts
used the term phyllids for these. Schuster writes of 'polymorphous
leaves', at the same time referring to individual units as 'leaf segments'.
The important points are (1) that they are cylindrical structures
three to five cells thick, and (2) that the whole relationship between
appendages and axis is an extremely plastic one, quite unlike anything
obtaining in 'leafy liverworts' generally; for they can arise singly,
in twos, threes or fours. They can hardly be regarded as leaves in
any ordinary sense and Schuster rightly emphasises that their
ontogeny needs further study. In this genus, as in *Haplomitrium,*
the superficial resemblance to a genus of Jungermanniales can be
considerable, as evidenced by the fact that a second species of
Takakia (*T. ceratophylla*) from the Himalaya was originally assigned
(over a hundred years ago) to *Lepidozia* (Grolle, 1963[154]). The
resemblance is borne out by the name, *T. lepidozioides,* of the
first-described species, which is currently known from Japan, North
Borneo and Western North America. Male plants of *Takakia* are
unknown.

4

THE SPOROPHYTE OF LIVERWORTS

Throughout the overwhelming majority of liverworts the sporophyte is made up of three components. These are the absorbing and anchoring foot, the stalk-like seta and the capsule which contains typically spores and elaters. Müller, like some others, uses the term sporogonium to refer to the capsule alone. In this text it is used in its wider meaning, when it becomes practically synonymous with the sporophyte. In a few liverworts, such as *Riccia,* the sporophyte is of a much simpler construction, whilst in *Anthoceros* it is different again in many important respects. Due consideration will be given to these two at a later point. More typical examples may be found in the Sphaerocarpales, Marchantiales, Calobryales, Metzgeriales and Jungermanniales. The sporophyte, small and compact by comparison with the thalloid or leafy gametophyte, has often been considered as wholly dependent. This is not completely true since chloroplasts occur freely in the capsule wall cells, commonly in the outermost two or three cell layers of the seta and occasionally (according to Müller, in *Sphaerocarpos*) in the foot.

The foot varies much in size and shape. Its width commonly exceeds that of the seta, so that it forms the swollen base of the sporophyte. It must surely serve as the principal pathway for absorption of nutrients from the gametophyte, although there is little experimental work to support this idea. The organ is nearly globose in many Marchantiales, including forms like *Corsinia* which lack a carpocephalum and others like *Reboulia* and *Conocephalum* which have one. In Metzgeriales and Jungermanniales it is often anchor-shaped in longitudinal section and the dagger-like apex may penetrate far into gametophyte tissue (Fig. 8B). Both *Sphaerocarpos* and *Monocarpus* have the foot quite well developed. On the other hand in some advanced Jungermanniales, e.g. *Lejeunea*, it is remarkably small.

The very young seta is composed of transversely elongated cells. The ultimate elongation of these cells in *Pellia* is said to proceed at a rate of 1 mm per hour and they may achieve individual lengths of 700 to 900 μ. A long seta is usual in Metzgeriales and Jungermanniales, even if few genera attain the great lengths (*c.* 5 cm) seen in *Pellia*. From Marchantiales, *Monoclea* differs sharply in its long seta (Fig. 3D). Obviously the value of a long seta is lost when that organ does not elevate the spore-container, but merely brings it nearer to the ground. This is what happens in the pendent sporogonia of Marchantiales with carpocephala, and the role of the seta is then discharged when it has carried the capsule clear of the 'head' itself and any attendant enveloping structures. Perhaps it is significant that even in *Targionia* and *Corsinia,* where no carpocephalum is formed, the seta is still short. The seta of liverworts is always a delicate and (after lengthening) an ephemeral structure; it is always nearly colourless at maturity and composed mainly of thin-walled cells. In the precise arrangement of these cells in a transverse section we have a potentially useful taxonomic character. In many genera the seta is eight to ten cells in diameter. In *Cephalozia* there is a central group of four surrounded by an outer ring of eight; whilst in *Cephaloziella* (according to Schuster) four minute central cells are surrounded by only four very much larger ones.

The form of the capsule varies from shortly cylindrical through ovoid to subspherical (Fig. 10). Its dark colour it owes to its densely packed contents; a gloss is imparted by a kind of cuticle, for it is important that there should not be excessive loss of water through the capsule wall before dehiscence. The variation in shape is of some taxonomic significance. Thus, the capsule is much more nearly ovoid-cylindrical in *Riccardia* and *Blasia* than it is in *Pellia* and *Fossombronia,* where it is nearly spherical. In the Jungermanniales ovoid capsules are common, but they are almost spherical throughout the sub-order Porellinae (*Frullania, Lejeunea,* etc.) whilst in *Geocalyx, Calypogeia* and a few others they are shortly cylindrical. The name *Sphaerocarpos* alludes to the nearly spherical capsule and the Marchantiales as a whole do not depart far from this shape. *Monoclea* and *Haplomitrium,* by contrast, both have elongated capsules (Figs. 3, 26). Liverwort capsules tend to be small compared with those of some mosses. One might expect capsules of carpocephalous Marchantiales (where several commonly hang from one 'umbrella') to be among the smallest, but this does not appear to be so. In fact, a capsule diameter of 1 to 1·25mm prevails among these, whereas figures as low as 0·5 to 0·6mm are found among Jungermanniales. A diameter of 1·5mm may be found in *Pellia* and *Riccardia* in the Metzgeriales.

The ripe capsule contains spores and elaters. The latter may be fixed individually to the capsule valves (cf. types of dehiscence described later) or they may be free among the spores. In *Pellia*, besides the free elaters, there is an elaterophore which bears a tuft of fixed elaters. This is at the base of the capsule (Fig. 10B). A comparable structure hangs from the apex in *Riccardia*. O'Hanlon[313] noted that in *Marchantia polymorpha* one elater was produced for every 128 spores. As Müller points out, the disparity is seldom so great as this. The elaters, long, narrow cells with single or double (occasionally multiple) spirals of thickening on the inside of the cell wall, are highly characteristic of liverworts. Also, as we shall see, they are highly efficient, in several different ways. In *Sphaerocarpos* and its ally *Riella,* in *Monocarpus* and again in *Corsinia,* no elaters are found and the spores are associated with rounded sterile cells which appear to have a nutritive function. These are the equivalent of sterilised spore mother cells, not modified elaters. An elater is normally formed from the equivalent of a row of spore mother cells. Horikawa and Miyoshi[192] have prepared a comparative survey of elater morphology and they reveal a surprising diversity. *Marchantia* and *Frullania* illustrate two extremes. *Marchantia* has exceptionally long elaters with tapered ends and double spiral thickening; *Frullania* is much shorter, flat-ended and has only a single spiral (Fig. 10L). In many Metzgeriales elaters tend to be short and broad, with rather imperfect development of the spiral thickenings. These authors conclude that the elater can be an important taxonomic character.

The wall of the capsule is one layer of cells thick in the Marchantiales, two to several layers in Metzgeriales and Jungermanniales. Characteristic bands of thickening run anticlinally across the wall cells in Marchantiales, and across the cells of one or more of the layers in the others. In *Lophocolea,* for example, the capsule wall consists of several cell layers, with thickenings in all of them, although the outermost layer has the largest and most heavily thickened cells. The cell-wall thickenings of the different layers present great diversity of detail, and in a taxonomically difficult group such as Jungermanniales, this detail (cf. Schuster[384]) may well command increasing attention in the future. It does however demand both skilled technique and precise description. The number of cell layers and the type and distribution of thickenings can often be

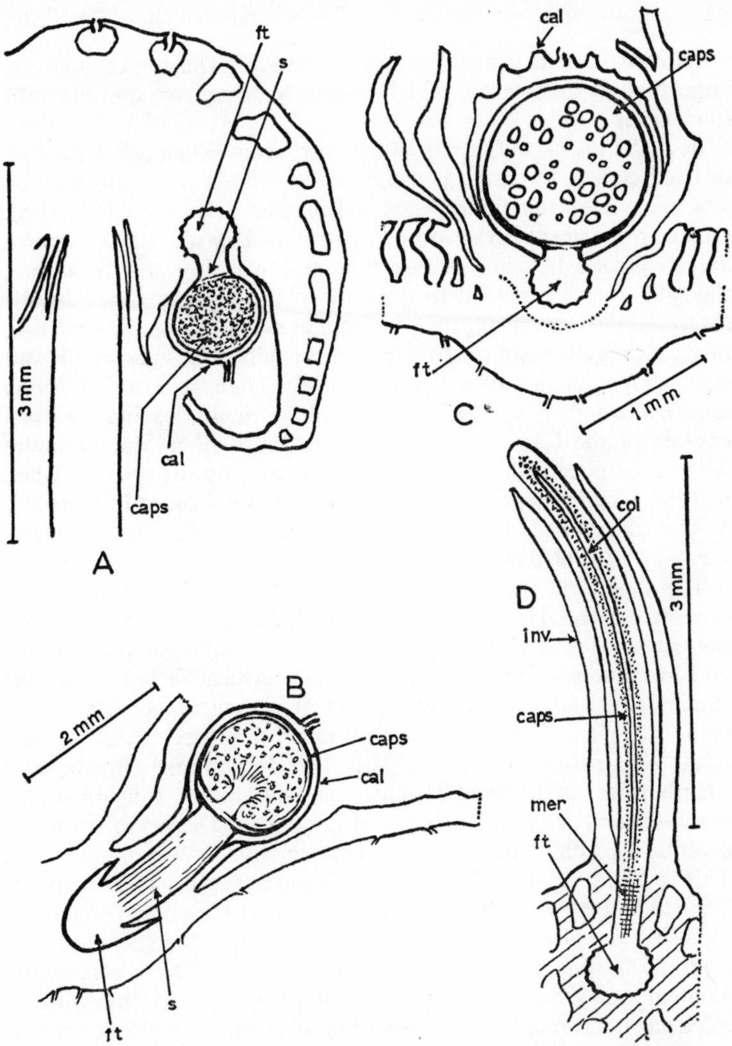

Fig. 8

Comparison of liverwort sporophytes, all seen in longitudinal section A. Part of carpocephalum of *Reboulia hemisphaerica* (itself in L.S.), with pendent sporophyte, still encased in calyptra. Chlorophyllose chambers of carpocephalum are indicated diagrammatically. B. *Pellia epiphylla*, showing elaterophore at base of capsule and remains of archegonial neck at extremity of calyptra; involucre (above) and extending thallus (below) have been cut away. C. *Corsinia coriandrina*. Note small, spherical foot, extremely short seta and capsule which contains spores and sterile nutritive cells. D. *Anthoceros husnotii*—with no seta, and capsule long, cylindrical (young in this specimen). cal. calyptra; caps. capsule; col. columella; ft. foot; inv. involucre; mer. meristem; s. seta.

characteristic of a genus, or a whole family. The thickenings are important in dehiscence, which normally occurs by splitting into four valves.

In *Pellia* the capsule splits to the base, from which arises the tuft of fixed elaters. In the related *Riccardia* fixed elaters stand more or less erect near the tip of each valve after dehiscence (Fig. 10D). Other arrangements exist in the four-valved capsules of *Lophocolea* and *Frullania*. In some Jungermanniales, e.g. *Lejeunea*, *Frullania*, the split does not extend to the base of the capsule. *Cryptothallus* (Metzgeriales) is very unusual, and reminiscent of the moss *Andreaea*, in its four valves failing to separate distally. Capsules of the old group 'Operculatae' of the Marchantiales (*Plagiochasma*, *Reboulia* and *Grimaldia*) shed a definite 'operculum' or lid, leaving the open capsule in the form of a neat urn. In most other Marchantiales with a carpocephalum the capsule breaks up irregularly when mature. Nor is there any definite mode of dehiscence in *Corsinia*. *Sphaerocarpos* and *Riella* are described by Müller as cleistocarpous (i.e. with no true dehiscence).

The spores of liverworts vary greatly in size, shape and surface ornamentation. Those of many Marchantiales and Metzgeriales present striking 'sculpturing' of the exospore (areolate, cristate etc.) and in *Fossombronia* this detail becomes critical in separating the species (Fig. 15G, p. 90). Throughout the Jungermanniales spores tend to be small, spherical and little ornamented. Schuster sees small size, in this instance, as derivative. In contrast with those of Marchantiae and Anthocerotae, the spores of Jungermanniae separate early from one another in the tetrad. The result (with few exceptions), is an 'apolar' spore—i.e. one with no 'triradiate mark'. This is quite an important morphological feature of the group. The student must not forget, of course, that the spore, despite its inclusion for obvious reasons in this account of the sporophyte, is itself the starting point of the new haploid (gametophyte) generation. Contemporary studies, utilising both scanning and transmission electron microscopy, are enhancing detailed knowledge of the spore wall.

Spore size and output per capsule often vary inversely. Thus, only about 200 of the massive tetrads of *Sphaerocarpos* are formed in one capsule. According to Jack[216] some 4,500 of the very large spores of *Pellia epiphylla* occur in one capsule. Müller[300] gives figures ranging from 2,000 to 8,000 spores per capsule for various genera of the carpocephalous Marchantiales. All these outputs are low compared with the figures attributed by him to some members of the leafy Jungermanniales. It is easier to credit the figure of about 24,000 cited for *Lophocolea cuspidata* than the estimated 400,000 for *Diplophyllum albicans* and 1,000,000 for *Scapania undulata*. These are high figures, even allowing for the small size of some spores in the Jungermanniales (some have a diameter less than one-twentieth that of the larger spores in the Marchantiales). In spite of a relatively modest output per capsule, the great fecundity of a plant such as *Marchantia polymorpha,* where many capsules arise in each of the numerous carpocephala, enables a single well-grown gametophyte to account for a prodigious spore production. O'Hanlon[313] estimated the figure of *c.* 7,000,000 from one plant (twenty-four capsules, each with 300,000). Precocious germination of spores is usual in *Pellia,* where they are often multicellular long before the seta has lengthened. This feature is also found in *Porella, Radula* and certain other genera.

It has been possible to examine sporophyte structure in outline only. The student would do well to compare this general account with that of moss sporophytes in Chapter 6. Superficially gametophytes are so much more diverse that one always thinks first of their characters when separating genera or families. A clear-cut difference, however, between two sporophytes could well be of great significance, for the liverwort sporophyte on the whole is a conservative structure. Thus, the difference between unistratose and multistratose capsule walls may well point to an ancient and fundamental cleavage between the two groups concerned. The same is true of differences in seta construction and of exceptional manifestations in the foot, such as the remarkable collar of rhizoid-like extensions found in *Schistochila*. The combination of massive sporophyte and single-layered capsule wall in *Haplomitrium* (Calobryales) is especially noteworthy. However, the subject of capsule wall structure is a large one that even now does not invariably receive all the attention it merits. The many variants occurring within the Jungermanniales are accorded full treatment by Schuster whose text (and excellent figures) should be consulted for further details. Among other things he is able to warn us that, in certain families at least, slight variations in the number of wall layers may be of no taxonomic significance.

Before we turn to two interesting aberrant cases, *Riccia* and *Anthoceros,* three general topics call for discussion. These are (1) embryology and early development; (2) protection and nutrition of the sporophyte; (3) spore discharge. The first played a big part in the older morphologists' discussions, but after some sixty years of intensive effort it was realised that embryological investigation was less rewarding than had been hoped. The lead had come from zoologists who from early times had been accustomed to draw on embryology for much significant evidence. Perhaps such evidence is to be found more readily in the unfolding of the complex organs and tissue systems of a highly organised animal than in the comparatively simple structure of a liverwort sporophyte. Nevertheless, two discoveries of some value were made. The first indicated that the early embryo either (1) passed through the octant stage (ball of eight cells) or (2) passed instead through a linear stage consisting of a chain of four cells (tiered embryo). The second discovery concerned a later stage, namely the origin of the sporogenous tissue. After the first periclinal walls have appeared it is evident that the forerunners of spore mother cells (archesporium) can be cut out either internal to or external to these walls. Putting the matter another way, we can describe the sporogenous tissue as referable to endothecium, or amphithecium. These two considerations were formerly judged of paramount importance and they weighed heavily with Cavers[74] in his very thorough discussion of the interrelationships of the Bryophyta. One hears much less of them today. Müller however, with long experience behind him, was careful to stress the importance of sporophyte characters in general, and those of development in particular.

The difficulty about early embryology in the present context lies in the fact that, whereas an octant stage is typical of Marchantiales and a linear stage typical of Jungermanniales, within the single close-knit group of the carpocephalous Marchantiales we find both represented. With *Conocephalum* showing the tiered, *Marchantia* the octant, arrangement of cells in the early embryo, it is difficult to regard such a difference as fundamental. The second criterion can be even less helpful since all liverworts except the Anthocerotae form their sporogenous tissue from the endothecium. We shall see that even in some Anthocerotae the position is equivocal.

Adequate protection and nutrition are very important for the success of the sporophyte and it is interesting to enquire how these are achieved in different instances. In such widely separated genera as *Sphaerocarpos, Corsinia, Targionia* and *Pellia* a similar general situation prevails. The green thalloid gametophyte houses the foot in each case and from its own resources provides for the adequate nutrition of the whole sporophyte. In all typical instances immediate protection for the sporophyte is provided by the calyptra, an enlargement of archegonial venter tissue; but external to this, each genus cited has its own additional protective structures. *Sphaerocarpos* has the prominent pear-shaped involucres which cover much of the thallus; *Corsinia* has a green curtain of tissue which arises late among the archegonia and which some have seen as the forerunner of a carpocephalum. The paired purplish black scales which enfold the sporophyte provide the characteristic involucre of *Targionia,* whilst in *Pellia* we meet with green involucres that are flap-like or collar-shaped according to the species. *Riccardia,* different again, has a particularly massive calyptra (six to eight cells thick), and lacks additional protection.

Quite another situation obtains in the carpocephalous Marchantiales, where the foot is embedded, not in thallus tissue, but in the substance of the mushroom-like head that forms the body of the carpocephalum. One notes that in the massive, rich green heads of *Reboulia hemisphaerica* (Fig. 8A) there is not only ample provision for photosynthesis (air chambers, green cells and barrel-shaped pores) but there is abundant storage tissue. Thus in the well-developed carpocephalum the inverted sporophytes are firmly embedded and well supplied. Additional protection external to the calyptra varies from genus to genus, often consisting of diaphanous sheaths of tissue. *Marchantia* represents the extreme case, with calyptra, perianth and involucre. *Fimbriaria,* allied to *Reboulia,* shows like *Marchantia* an individual perianth around each archegonium; and this perianth becoming laciniate into four to sixteen strips, looks like a Chinese lantern around the mature sporophyte. Sometimes the strips remain coherent at their tips and the spores are shed through the slits so formed.

The Jungermanniales, bearing their archegonia at the tips of slender leafy shoots, would seem less favourably constructed for the protection of the ensuing sporophyte. Almost all, however, have the tubular or funnel-shaped perianth that has been described in Chapter 3. It is reduced or lacking, however, in genera where protection for the developing sporophyte is provided by special modifications of axial tissue in the general region of the archegonial insertion. This is a most interesting adaptive development which in its extreme form results in the sporophyte coming to be housed in a pendulous pouch, or marsupium, of gametophyte origin. Just as a calyptra implies a potentiality of further growth in the tissue of the archegonial venter, so a marsupium implies such a potentiality in a certain part of the stem itself. Precisely what part of the stem is affected, the exact pattern of growth and the morphology of the resulting structure all vary from genus to genus. Considered in detail the subject is complex. We can give only an outline.

The principal kinds of marsupium were clearly set forth by Cavers,[74] who also pointed out that an essential feature was the modification or loss of true calyptra, perianth or both. Later on Knapp[236] carried out a monumental survey of the subject and it was he who introduced the term 'shoot calyptra' (Sproszkalyptra) for those cases in which proliferation of axial tissue resulted in an apparent calyptra that was composed of extended receptacular tissue, the unfertilised archegonia being carried far up in the process. Recently, Schuster[384] has provided a thorough treatment of the subject. His excellent illustrations repay careful study. Schuster, however, draws a subtle distinction between a 'coelocaule' in which the main feature is deep penetration of gametophyte axis by the sporophyte and a 'shoot-calyptra' in which the emphasis is on the upward growth of receptacular tissue. To all the more extreme types of structure he applies the term 'perigynium'. This term underlines the fact that we are dealing here with structural specialisations surrounding the 'gynoecium'—i.e. at first the archegonia, later the sporophyte. We can see such structures as primarily protective, and to a certain extent nutritive.

It is the general biological interest, rather than the slightly confused terminology, that concerns us most in the present context. Here is a unique series of structural modifications for sporophyte protection and nutrition. As illustrative examples we can take only three. Our first is *Trichocolea tomentella* (Fig. 9D). This is, strictly, a 'coelocaule'. Knapp[236] showed that two elements were involved here, (1) the boring down of sporophyte into gametophyte and (2) a tubular extension of stem tissue from well below archegonial level. It betrays its origin by the leaves which it bears and the bracts that crown it. The final result is to enclose the whole developing sporophyte in a massive, club-shaped extension of the shoot apex. Both true calyptra and perianth are missing. Our second example is *Nardia geoscyphus* (Fig. 9C) which represents a modification of what is sometimes known as the '*Isotachis* type'. The essential feature is a stout, ring-like development of perigynium which comes from tissue surrounding the archegonial group. It is asymmetrical, rhizoid-bearing and clearly in part a storage unit. Unfertilised archegonia are not carried up and one notes that a genuine calyptra and a vestigial perianth survive.

Our third example is the one to which the term marsupium is most truly applicable. It is found in *Calypogeia* (Fig. 9A,B,E). It is a more extreme development of the last. The pendulous, almost tuber-like structure is well equipped with rhizoids externally, whilst mucilage hairs line the cavity above. A short calyptra survives but a perianth is rendered quite superfluous. Extensive adjustments have taken place in axial tissues to achieve this effect, which must represent the most perfect development to be found among liverworts for the protection and nutrition of the sporophyte.

It is not intended to dwell at length on capsule ripening, with its attendant processes of tetrad division in the spore mother cells, elater maturation and deposition of thickenings in the ripening capsule wall. Dehiscence of the capsule too has already been considered in a general way, and we may turn now to examine in

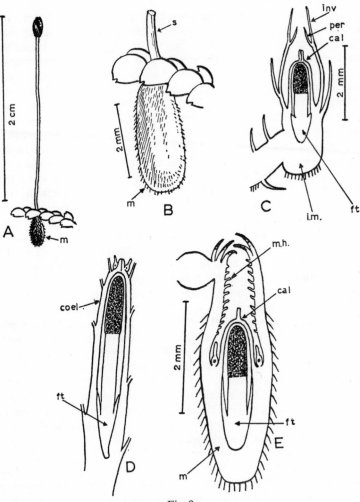

Fig. 9

A. *Calypogeia* sp. Part of leafy shoot, with sporophyte and conspicuous marsupium. B. Enlarged view of the massive, pendulous marsupium. C–E. Diagrammatic longitudinal sections of 3 sporophytes to show different types of marsupial structure (re-drawn from Cavers). C. *Nardia geoscyphus* type. D. A type found in the genus *Trichocolea*. E. *Calypogeia* type (at younger stage than A–B) cal. calyptra; coel. 'coelocaule'—formed of stem tissue—and a kind of marsupial structure, although loosely called calyptra in descriptive works; ft. foot; i.m. incipient marsupium; inv. involucre; m. marsupium; m.h. mucilage hairs; per. perianth; s. base of seta. In C–E the capsule is shown dark. Note that in D unfertilised archegonia are carried up, in E they stay down at capsule level. For further explanation see text.

some detail a few representative methods of spore discharge. We are fortunate in having available Ingold's extremely clear and thorough exposition on this subject.[203] and it should be consulted by everyone interested in the precise role of the elaters in different examples. Following up the earlier observations of Kamerling[227] and amplifying them with his own, Ingold recognises three distinct mechanisms effecting spore discharge among liverworts. These are (1) the water rupture mechanism, seen in *Cephalozia, Lophocolea* and most leafy liverworts; (2) the hygroscopic mechanism, seen in *Marchantia* and indeed most members of the Marchantiales with elaters, also in *Pellia* and other Metzgeriales; (3) the spiral-spring mechanism which appears to be peculiar to the Frullaniaceae and Lejeuneaceae. The first and the third are violent methods, the second much less so. In all of them the elaters are the operative structures, together with the opening valves of the capsule wall. The essential feature of *Cephalozia,* and liverworts like it, lies in the great strength of the bi-spiral thickening on the inside wall of the clear, water-filled elater (Fig. 10H). On exposure to dry air this double spiral band contracts sharply, bringing the contained water under severe tension. Finally, the tendency of the spiral to revert to its original shape is too strong for the cohesive power of the reduced water content of the elater; the water 'breaks' and a gas phase takes its place. Instantaneously, the elater untwists and, as Ingold graphically describes it, this 'tears the elater free from the sporangium wall, hurls it still rotating into the air, and flings off the attached spores'.

By comparison, the spiral bands in *Marchantia* and *Pellia* are weaker. According to Horikawa and Miyoshi,[192] *Marchantia* shows a double spiral, *Pellia* two to six spiral bands within the elater. In these cases the spirals never reach a stage, on withdrawal of water from the elater, when the strength of their tendency to untwist is sufficient to induce water rupture. As a result, they perform only comparatively feeble twisting or 'wriggling' movements in response to changes in atmospheric humidity. By such movements, of course, it is possible to 'fluff up' the mass of spores and facilitate their gradual dispersal.

The special arrangement in Frullaniaceae and Lejeuneaceae depends upon their having a series of elaters extending from the roof to the floor of the globose capsule, and firmly attached at both ends. Hence, as the four valves of the capsule bend back in dehiscence, there is a fraction of a second when each elater is drawn out so that its single band of thickening forms a stretched spiral spring in a water-filled tube. Ingold emphasises that this stage is passed through rapidly, for in less than a second after the onset of capsule

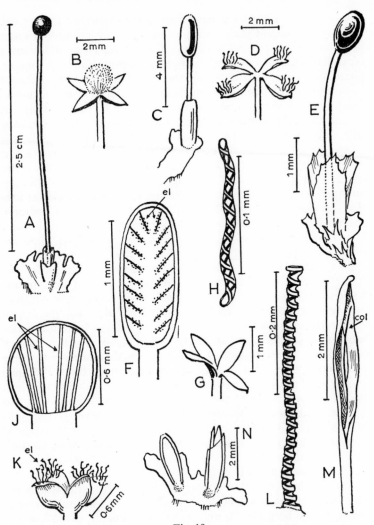

Fig. 10

Dehiscence of capsule and spore discharge in liverworts. A–B. *Pellia epiphylla*, capsules before and after dehiscence. C–D. *Ricardia pinguis*, the same. E–H. *Lophocolea heterophylla*. E. Capsule, seta and perianth. F. L.S. intact capsule, diagrammatic (after Ingold). G. Old, empty capsule. H. Elater; note double spiral. J–L. *Frullania dilatata*. J. L.S. intact capsule, diagrammatic (after Ingold). K. Old capsule, after spore discharge. L. Elater; note single spiral. M. *Anthoceros* sp. Distal part of capsule in dehiscence. N. *Notothylas orbicularis*, part of thallus with two capsules, one beginning to open. col. columella; el. elaters. For further explanation see text.

opening all the elaters will have been torn free at their lower ends. Quickly the base of each elater moves out in an arc until a series of elaters—their original shape resumed—stand more or less erect on each of the four expanded valves of the capsule wall (Fig. 10J,K,L). Spore discharge is effected by this rapid upward and outward movement of the freed bases of the elaters. Any subsequent movement that they perform, light twisting induced by further drying, is unimportant. This unique method, it will be observed, is associated with a unique internal organisation of the capsule. That fact did not escape the attention of the earlier observers who also noted that in these families the splitting of the capsule wall into four valves only extends for two-thirds of its length, leaving the basal third as an intact bowl. Spruce[407] attached such importance to these features that he divided the whole range of liverworts, which compose the modern Metzgeriales and Jungermanniales, into two groups, one containing *Frullania, Lejeunea* and their immediate allies, the other all the rest.

So far this chapter has been descriptive rather than phyletic in its approach. One reason is that two very important links in any phyletic series remain to be considered. Each is very different from any hepatic sporophyte so far examined. Of the two, *Riccia* and *Anthoceros,* the former will be described first. It stands apart because it is the simplest known sporophyte structure among bryophytes. In all species of *Riccia* a capsule alone exists, there being no trace of seta or foot. At an early stage the archegonial venter will have become two-layered by periclinal divisions, whilst the nearly spherical capsule consists of a single-layered wall and the enclosed mass of sporogenous tissue. Sometimes certain cells of this mass fail to form spores; such sterile cells, or 'nurse cells' (from their undoubted nutritive function), were detected by Pagan[315] in *R. crystallina.* At maturity the capsule wall ('jacket layer' of Smith[403]) breaks down to a variable extent. As the inner layer of the old archegonial venter commonly breaks down too, the spores will ultimately lie free in a delicate sac formed only of the outer layer of the venter. As Smith points out, we have the anomaly of the so-called sporophyte of *Riccia* at this stage consisting of a sac provided by an earlier gametophyte housing spores that represent the new gametophyte—and no diploid tissue at all! Spore dispersal awaits the break-up of the surrounding thallus tissue, but this does not mean, of course, the death of the plant. Sometimes one can see the neat rosettes of living *Riccia* species, with the tips of the thallus a fresh, vivid green, the older parts duller, erumpent and spattered with loose spores.

The embryology of *Riccia* seems to be of little significance since both tiered and octant types are known within the genus. *Oxymitra,* closely allied to *Riccia,* to a large extent shares its simplicity of sporophyte, although it is normal here for a number of sterile cells to occur within the capsule. So far as sporophyte structure is concerned, neither genus of the family Ricciaceae helps much to bridge the gap between them and other Marchantiales. Lack of foot and seta effectively sets them apart.

As is well known, Bower[47] seized upon the extraordinarily simplified sporophyte of *Riccia* as the natural starting point in the evolution of land plants through a process of sterilisation of more and more 'potentially sporogenous tissue'. Seen in this way, as the base-line of his antithetic theory of the origin of land plants from algal ancestors, *Riccia* acquires great importance morphologically; and from it can be traced a beautiful ascending series through *Corsinia* to the higher Marchantiales. Cavers[74] accepted without question the validity of such an ascending series. Many bryologists today, however, would regard the very simple sporophyte of *Riccia* as the product of reduction. Müller and Evans are two of great eminence who in recent years have shared this view.

The sporophyte of *Anthoceros* is of the greatest interest, and its many unusual features have given support to the argument for raising the order Anthocerotales to the rank of a class equated with mosses and hepatics. It will be helpful to summarise the most important features of the sporophyte, and then elaborate on some of them. Thus, (1) there is no seta, only a massive foot and a long, cylindrical capsule; (2) the capsule grows for a long period, from a basal intercalary meristem (Fig. 8D); (3) there is elaborate internal organisation into central columella, elongate but domed sporogenous tissue of amphithecial origin, and wall composed of several layers of green cells and an epidermis with stomata (Fig. 25, p. 177); (4) dehiscence is by two (at times more) valves, the rupture extending downwards from near the apex, thus continually exposing spores of more and more recent origin. (5) The spores are intermixed with 'pseudo-elaters' which, despite lack of spiral thickenings, perform some hygroscopic movements which aid in dispersal. Proskauer[343] has given a very thorough account of this and the associated spiralling of the dried-out valves.

It is plain that the long-growing capsule, the massive foot and the complex internal organisation, with photosynthetic power unmatched elsewhere among liverwort sporophytes, combine to give this unique structure almost all that it needs to become free-living. Normally, however, the foot is firmly anchored in gametophyte tissue; and the slender light green capsules, which remind one of the seedling shoots of a fine-leaved grass, are semi-dependent.

Many years ago, Campbell[65] placed on record an isolated instance of this sporophyte achieving something more. He found unusually large plants of the Californian *Anthoceros fusiformis* which had grown for nine months. They were on the brink of achieving independence from the aging gametophyte. Some parts near the base of the capsule had ceased to produce spores almost completely, and there were corresponding increases in the extent of photosynthetic tissue, in size of obviously conducting columella and massive, absorbing foot. Campbell's sporogonia came nearer to full independence than anything ever seen before in a bryophyte.

Much of the interest aroused by Campbell's discovery hinged on the fact that *Anthoceros* was seen at that time (by most bryologists) as representing the peak of advance in liverwort evolution. Not only did it display a series of unique features (summarised above) but it was clearly viewed as 'on the upgrade' towards the achievement of pteridophyte status. Now this status implies the existence of true vascular tissues and a diploid plant body which, even at its simplest, must be much more than a solitary spore-container borne up on a pedestal which is itself firmly anchored in the gametophyte. Nearly fifty years have passed since those 'outsize' specimens of *Anthoceros fusiformis* were found and we are still as far as ever from bridging this huge gap between bryophyte and pteridophyte status. Also, alternative views on liverwort evolution as a whole have gained in popularity. The origin of *Anthoceros* remains obscure but it is no longer regarded as at the end of a long ascending series. It is worth bearing in mind, however, that the genus does not stand alone. Apart from 'splits' off the large genus *Anthoceros* itself (*Aspiromitus, Phaeoceros*) there are *Notothylas, Dendroceros* and *Megaceros,* all of which are referable to the Anthocerotae. Of these, *Notothylas* may be examined in the present context.

The short capsules of *Notothylas orbicularis* (only *c.* 2 mm long) grow out nearly horizontally from the fertile branches of the rosette (Fig. 10N) and seem remarkably unlike those of *Anthoceros* species. Not only are they short and compact, but chlorophyll is almost absent and there are no stomata. Early investigators noted the variable extent of a sterile columella, which in some species appeared to be totally lacking, so that the whole endothecium formed spores and elaters. Lang,[247] in a careful and thorough study of *N. breutelii,* found that the central products of the basal meristem at first formed sporogenous tissue but later in the life of the capsule gave rise to sterile cells which could result in a short columella being present in the mature capsule. He also demonstrated that in the embryology of the sporophyte *Notothylas* was perfectly in line with the prevailing pattern of the Anthocerotales. More recently Kashyap and Dutt[228] and Pandé[317-18] have studied Indian species of the genus and confirmed that the columella is highly variable from species to species. Elaters in *Notothylas* commonly have at least rudimentary rings or spirals and dehiscence is usually by the imperfect separation of two broad valves, from the apex downwards. Pandé cites species, however, in which the opening is follicular, along one suture only.

Notothylas at once appears as a link between *Anthoceros* and more normal liverworts; for it is evident that many of the most arresting sporophyte characters of *Anthoceros* are either lacking or present in modified form here. Cavers[74] clearly regarded it as more primitive than *Anthoceros* and even went so far as to name *Anthoceros hallii* as a link (in the upward march of events) between the two genera. It is interesting that in 1907 Lang had put forward the opposite view, that *Notothylas* was a reduced member of the Anthocerotales that had largely lost its columella. If one agrees with Lang in this one can continue to regard the sporophyte of *Notothylas* as a 'link' with those of ordinary liverworts only if one begins the series with *Anthoceros* and regards Jungermanniales and Metzgeriales as examples of still further reduction. In my view, this involves considerable difficulties. Perhaps it is merely a reduced derivative from 'Anthocerotalean' stock which is secondarily modified in various ways that are directly connected with its small size and ephemeral life. The basic aberrant characters (e.g. chloroplast structure, form of sporophyte, etc.) are still present and even the 'elaters' are very different from those of most ordinary Hepaticae. The weak rings or spirals may have been independently acquired and there is no compelling reason why it should be seen as a link between Anthocerotae and Hepaticae in either the upgrade or downgrade direction.

We shall return in the final chapter to look again at these exceptional sporophytes in the wider context of liverwort evolution as a whole. Meantime, they stand at opposite poles in this matter of structural complexity. Although there have been many recent papers dealing with either *Riccia* or *Anthoceros,* these have been in the main descriptive, and few facts have appeared which could throw light on the wider question of relationships. Proskauer's report[346] of spiral thickenings in some columella cells of *Dendroceros crispus,* however, is one which may well prove significant.

5

GAMETOPHYTE OF MOSSES

The branched, filamentous protonema which forms on spore germination is invariably succeeded by a leafy shoot system or gametophore. In some mosses, for example *Sphagnum* and *Andreaea*, the forerunner of the leafy shoots is a plate of cells. The leafy shoots present a great diversity of gross form, and of fine structure, in the different families and genera of mosses. They offer many taxonomic characters of importance; yet the gametophyte has not been much called upon to provide evidence regarding moss evolution. There exists for mosses no modern review of interrelationships comparable with those of Evans,[120] Fulford[129] and Schuster[384] for hepatics, in which the gametophyte could have received due weight in the discussion. This is not wholly surprising because for all its diversity the gametophyte of mosses shows an underlying uniformity of plan. Moreover, the existing differences are in the main of a kind to be accorded only minor importance in an evolutionary context.

With very few exceptions the leafy shoot grows by means of a tetrahedral apical cell, in the form of an inverted pyramid with three cutting faces. Even in such a case as *Fissidens*, where the apical cell is characteristically two-sided rather than three-sided, that in the youngest shoots has the normal three cutting faces. In all ordinary examples it follows therefore that the resulting stem bears leaves whose basic arrangement is in three ranks. Owing to growth torsions and consequent displacement, this precise arrangement is soon lost in most instances. *Fontinalis,* however, continues to display the strictly three ranked leaf arrangement. If the naturally twisted ('rope-like') shoots of *Grimmia funalis* be 'unwound' by hand the underlying three-ranked leaf arrangement is again plain. In mature shoots of *Fissidens* species the leaves are strictly in two ranks (cf. also pp. 70–1). The apparent five-ranked leaf formation of some species of *Sphagnum* (e.g. *S. quinquefarium*) is secondarily acquired, and the normal three-sided apical cell prevails in this genus too. Thus, stem and leaves, or cauloid and phylloids as some[238] would call them, are two of the fundamental organs of the moss gametophore. The third is the rhizoid system. This closely resembles protonema in structure and is developed to a varying extent.

Much of the characteristic appearance of particular mosses is imparted by their branching system. Ruhland[370] has indicated the manner of formation of branches. Unlike higher plants, many mosses form branches just below leaves rather than in their axils. The branch originates by a cell belonging to the surface layer of the stem becoming secondarily specialised as a new apical. It then proceeds to function like the original apical cell. If growth of the branch is rapid one may gain the impression of dichotomy (forking). If many new apicals arise regularly on opposite sides of the stem, each initiating a lateral of limited growth, a pinnate or 'plumose' type of shoot will result, such as one sees in *Ctenidium molluscum,* or even more strikingly in the boreal forest moss, *Ptilium crista-castrensis.* Thus, the frequency of new initials, together with the direction of growth of both main stem and lateral branches, will largely determine the 'habit' of a particular species. Where a main stem grows erect for several centimetres, bearing laterals only in the form of a subterminal crown of branches, a 'miniature tree' (or dendroid habit) results. This is well seen in *Climacium dendroides* (Fig. 22J), found by lake margins and in dune 'slacks', and to a lesser extent in *Thamnium alopecurum*, a plant of limestone woods and boulders in mountain ravines. Hörmann[193] has recently prepared a rather full account of the structure of these two dendroid mosses. We return to this topic in Chapter 10.

It will be well to examine at this stage the purely vegetative gametophyte structure in selected examples from each of the principal sub-classes (Reimers—see Chapter 1) of mosses. The following will be considered: *Sphagnum, Andreaea, Funaria, Hypnum, Buxbaumia* and *Polytrichum.*

These will serve to display the salient lines of specialisation in the leafy shoot as these are exemplified by major taxonomic groups. Descriptions of these examples are available in many textbooks and it will suffice to mention their principal features. *Sphagnum* offers the greatest number of unique features, and it will be considered first.

The genus *Sphagnum* is very large, with over 300 species, distributed in almost every part of the world. They are known as bog

C

mosses, and they make these wet habitats increasingly acid by their activity. On the erect-growing stems the branches arise in fascicles, often five together (Fig. 22D). Commonly some will grow out laterally, whilst others of the fascicle are 'descending branches' and hang down very close to the main stem. The branches near the apex are usually shorter, stouter and crowded. They compose the so-called comal tuft, but as new growth appears these in turn will be displaced downwards.

The main stem is comparatively stout (diameter up to 1·2 mm) and shows well marked differentiation of tissues (see Chapter 9 and Fig. 20F,G). The leaves on it are usually different in shape, size and details of cell structure from those on the branches. It is worth while to dissect out some young leaves from a bud of *Sphagnum* to see the regular manner in which cell division proceeds, with its resulting pattern of alternating living green cells and clear, hyaline, water-filled cells with ring-like thickenings. This has often been figured,[74,320,403] and requires emphasis here only because it is an arrangement of cells that sets *Sphagnum* apart from all other mosses (cf. Chapter 11). There is no midrib. Mature plants are without rhizoids. Great rigidity is not required of *Sphagnum* stems, for mostly they grow in the dense clumps familiar to everyone who walks much in moorland country. Except in dry weather, these clumps are charged with water (numerous holes providing direct entry to the hyaline leaf cells); and the plants glow with many bright colours—deep red in *S. rubellum,* salmon to rose-pink in *S. plumulosum,* orange-brown and various shades of green or yellow in many others. Paton and Goodman[324] found that the pigment responsible for the rose-red colour was an anthocyanin very firmly held in the cell walls.

Thus, in many structural features of the gametophyte *Sphagnum* stands apart; and we have yet to consider sex organs and sporophyte which will provide further evidence of isolation. Thus, it is not easy to concede to Gams' plea[137] that *Sphagnum* be viewed as a 'reduced' evolutionary line, derived from Leucobryoid ancestry. *Leucobryum* is an important genus of the order Dicranales in the Bryidae, which shares with *Sphagnum* the capacity to store water in certain dead, hyaline cells of the leaf. It is difficult to see that it shares much else. Even this leaf is of very different basic structure from that of *Sphagnum*. In it there is a layer of green cells sandwiched between two layers of hyaline cells (cf. Fig. 21E,F, p. 131). *Leucobryum* is much closer to the tropical genus *Octoblepharum,* which belongs to the same family (Leucobryaceae), and also has multistratose leaves. The gametophyte of *Sphagnum* offers the taxonomist a wealth of characters which present a reasonably constant pattern in any one species despite the considerable phenotypic variability which prevails in size, colour and general habit (cf. the important nomenclatural revision of European species of *Sphagnum* published by Isoviita[211] in 1966).

The small, blackish-green or olive-brown tufts of *Andreaea* present few features of the leafy shoot to set them apart from many Bryidae. Goebel[148] alluded to a fundamentally primitive mode of segmentation in the leaf of *Andreaea,* however. The leaves are nerved or nerveless, perichaetial leaves being greatly enlarged. Cells in both stem and leaves have notably thick walls. Normal rhizoids tend to be replaced by cylindrical or plate-like structures. Only this last is a really unusual feature and the relative isolation of the genus is based mainly on characters derived from sporophyte, sex organs and prothallus, probably in that order of importance. Such a plant as *Andreaea rupestris* looks not unlike small states of *Rhacomitrium heterostichum* var. *gracilescens,* which may be growing on the same siliceous mountain rocks. The plate-like protonema doubtless makes for firm anchorage on the hard rock. In common with *Sphagnum* (and in contrast with all ordinary mosses) a gametophyte stalk, or 'pseudopodium', replaces a seta functionally and lifts the capsule clear of the perichaetial leaves.

Among Bryidae, our two examples offer many contrasts. *Funaria* illustrates the erect, little-branched 'acrocarp' habit; *Hypnum cupressiforme* the prostrate, freely branched, 'pleurocarp' habit. The leaves in *F. hygrometrica* are large, oblong-ovate and shortly acute at the apex, with large cells of rectangular or hexagonal-rectangular outline in surface view. The leaves in *H. cupressiforme* are strongly curved, finely acuminate (cf. Fig.22G, p. 141), with long narrow cells throughout most of the leaf and groups of isodia-metric cells in its basal angles. A midrib is present in the leaf of *Funaria,* and the chloroplasts in the leaf cells are large and prominent. No midrib exists in *Hypnum cupressiforme* and the chaff-like leaves are noticeably poor in chloroplasts. Although *Hypnum cupressiforme* is the most notoriously polymorphic British moss, and one of the commonest, it never departs from this description in any essential feature. The differences concern size and habit in the main, though some observers (Guillamot,[160] Doignon[112]) have claimed small but constant differences in cell measurements and other fine details between the many so-called varieties of this species. Indeed there are those who raise some at least of these taxa to full specific status (cf. p. 152).

The above contrast hints at the breadth of gametophyte expression in the Bryidae, but it does not explore it. A dozen examples could be found, differing as widely again in every particular of leaf structure and each of distinctive habit; and many would be found to typify a whole family or sometimes an order of the subclass Bryidae. Confining himself to British mosses, the interested student would do well to look up descriptions, figures and, better still, specimens of the following illustrative examples: *Fissidens, Dicranum, Tortula, Grimmia, Bryum, Mnium, Bartramia, Ulota, Fontinalis, Neckera, Thuidium, Brachythecium* and *Rhytidiadelphus*. Among so many small but clear-cut differences of habit, leaf shape and 'pattern' of leaf cell structure it is often difficult to find any that are fundamental. Hence the Bryidae compose a single large, unwieldy subclass of the class Musci.

Buxbaumia, always a notable genus, and now made more famous as the emblem and title of the Dutch bryological journal, has a very remarkable gametophyte. Perhaps the clearest description and figures of it are those given by Goebel,[148] who described *Buxbaumia* male plants as 'about the simplest moss plants we know'. They could hardly be simpler, for they consist only of protonema which bears at the tips of some of its branches a curious hood-like 'leaf' protecting the antheridium. This 'leaf' is bizarre in form and quite minute. It is, however, chlorophyllose. In the slightly more elaborate female plant the protonema alone is green. Unfortunately both British species of *Buxbaumia* are very rare, for it is a genus of the greatest interest, not merely in the present context, but also for its sporophyte and mode of life. Growing always on substrata rich in organic matter, it seems to be in part saprophytic. Although Goebel is often quoted as having believed in widespread reduction among bryophytes, one notices that in the special case of *Buxbaumia* he concluded that it had '*stood still*' in a stage which other Musci had passed. In a word, it is an archaic form. We shall see later on how far a study of its sporophyte bears out this view.

Polytrichum presents a highly organised gametophyte. The genus is large (over 100 species) and of nearly world-wide distribution. Richards and Wallace[363] recognise eleven species as British, of which the largest, *P. commune*, may be taken as an example. The erect gametophores normally attain a height of 25–30cm in this species but, as in all typical members of the genus, they spring from a horizontal underground so-called rhizome. This organ bears abundant rhizoids which tend to be massed together and spirally twisted; they have been shown to be important in both external and internal conduction. The anatomy of the stem is complex. The role of a central strand in conduction is controversial (cf. Bowen,[45] Mägdefrau[278]) but whatever importance it may have it seems to be supplemented by an external pathway furnished by the widely sheathing leaf bases (cf. Chapter 9). These clasp the stem so closely as to provide a capillary channel (Fig. 20A). They can also make the stem appear glossy when dry—a useful 'field character' for *Polytrichum commune*.

Each leaf consists of the almost colourless sheathing base and a lanceolate 'limb', which in all species of *Polytrichum* appears relatively firm, opaque and dark green. A transverse section shows it to consist of a very wide nerve, flanked by narrow wings of tissue that are one cell thick. The wide nerve is many cells thick and includes both thick-walled and thin-walled elements (Fig.21G,H, p. 131). The firm, opaque character of the leaf arises from close-set rows of lamellae borne on the adaxial surface. Each lamella appears in section as a chain of green cells crowned by a cell of special form, often almost colourless and with thickened outer wall. This cell provides an important taxonomic character; it is normally bifid in *P. commune* but, according to Magnée[279] complete reliance cannot be placed on this. Clearly these lamellae, which in the intact leaf are like green walls running the length of the limb, and separated by extremely narrow spaces, are an important photosynthetic tissue. There are commonly thirty to fifty to a leaf, and they find their closest parallel in the photosynthetic tissue on the upper surface of the thallus in the liverwort *Riccia*. This short summary is enough to indicate in *Polytrichum* a shoot structure that is distinctive in many ways. We must turn to the close allies of *Buxbaumia* and of *Polytrichum* for links between them and more normal moss shoot structure.

Diphyscium foliosum is widespread on turfy banks in north and west Britain, so that material is not difficult to obtain. It possesses the disproportionately large sporophyte and diminutive gametophyte of *Buxbaumia,* and like that genus is dioecious. The male plants, however, are much larger than those of *Buxbaumia* and the female gametophore is a well-developed green leafy shoot. The strap-shaped leaves, much curled when dry, resemble those of some species of *Tortella* and *Trichostomum* among the Bryidae but differ in being two cell layers thick. The laciniate perichaetial bracts are peculiar and the sporophyte proclaims its affinities with that of *Buxbaumia*. Thus, *Diphyscium* provides some link between *Bauxbaumia* and the Bryidae. The close relatives of *Polytrichum* also provide a link between it and less specialised mosses, so far as gametophytes are concerned. Thus, in species of *Atrichum* the leaves appear filmy in character, with many resemblances to those of *Mnium* species. The nerve is narrow and the lamellae few in number. Again, examination of the sporophyte will leave us in no doubt as to their affinities.

Although gametophyte characters, especially those of the leaf, are so widely used in moss taxonomy, we must conclude that for displaying underlying affinities between two plants real resemblances between the two sporophytes are normally essential. The genus

Aloina (Pottiales), for example, shows a system of lamellae on the adaxial surface of the leaf somewhat like that of *Polytrichum,* but the merest glance at the sporophyte shows that the two genera are quite unrelated. The above review suggests that only to a certain degree does each subclass have its own characteristic type of gametophyte. In any case, the genera of Bryidae so far outnumber those of all the other four subclasses together that it is to the diversity of leafy shoots found among these that attention must now be turned.

Exceedingly small moss gametophytes are not uncommon, the tiny shoots consisting of little more than a few leaves surrounding the groups of sex organs. An extreme case in the British flora is *Ephemerum,* with *E. serratum* its commonest species (Fig. 13H). Here the protonema is long-persistent and the leafy shoots are almost microscopic. Still more extreme is *Ephemeropsis,* with two species, one known from Java, the other from New Zealand and Tasmania. In this genus all leaves are directly associated with sex organs and are only about 0·2 mm long. The plants grow on the surfaces of leaves or twigs and, as Sainsbury[372] points out, without sporophytes the presence of the moss is scarcely detectable, so close is its superficial resemblance to a filamentous green alga. Such examples as this make the gametophyte of *Buxbaumia* appear less isolated. Nor must we forget the surprisingly small (dwarf) male plants that are found in some species of *Dicranum* and certain other genera.

The two most marked modifications of leaf arrangement known are (1) the strictly two-ranked and (2) the strictly three-ranked arrangement, in which one rank is composed of much smaller leaves. A tendency for leafy shoots to become flattened in one plane is not uncommon, being seen in *Plagiothecium, Isopterygium* (Fig. 12C), *Neckera, Homalia* and some others; but in none of these are the leaves strictly in two ranks, and the usual three-sided apical cell is present. In the enormous genus *Fissidens* (Fig. 12B), by contrast, there are two ranks only and an apical cell with two cutting faces functions for most of the life of the plant. *Fissidens* is odd in another respect, namely in having leaves that are boat-shaped, with an additional wing of tissue. In typical members of the genus this 'wing' is so extensive as to alter completely the appearance of the leaf; and the midrib runs out into it. A link with more normal leaf structure is seen in the related genus *Sorapilla,* where only a small dorsal wing is formed. Strictly two-ranked leaves occur in a few other instances, e.g. in *Distichium* which is thus easily distinguished from other Ditrichaceae, and in scattered examples among non-British genera.

Apart from *Fontinalis* (already mentioned), a symmetrically three-ranked leaf arrangement is found in *Tristichium, Triquetrella* and other genera of widely scattered affinities. More remarkable is the situation where the ventral rank of leaves has become reduced in size and the term 'amphigastria' has been borrowed from liver-worts to describe them. This is seen well in the family Hypoptery-giaceae (Fig. 11A), with few genera and under 100 species altogether.

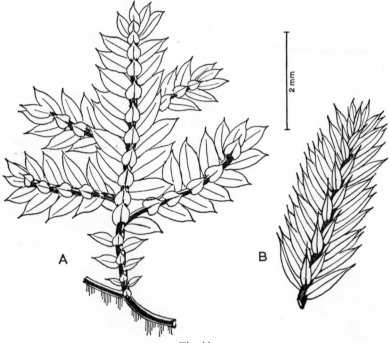

Fig. 11

Unusual moss shoots, with leaves in ranks and of different sizes. A. *Hypopterygium* sp., ventral view of frond-like shoot system. B. *Rhacopilum* sp., dorsal view of single branch. Drawn from specimens, the former adventive in a Reading conservatory, the latter from Idanre, Nigeria. Both are to same scale.

Catharomnion is monotypic and *C. ciliatum,* from New Zealand and Tasmania, is also remarkable for its long-ciliate leaf margins. *Hypopterygium* is by far the largest genus and is widely distributed in the tropics and sub-tropics, extending south to New Zealand and Tasmania. Sometimes a species of *Hypopterygium* may establish itself in suitably shaded artificial habitats in British botanic gardens. Goebel[148] referred to the anisophylly in this genus and indicated its probable derivation from a radially symmetrical shoot structure.

In the large tropical genus *Rhacopilum* one finds the diminutive leaves inserted along the dorsal face of the axis, their points deflected alternately right and left (Fig. 11B). The ventral surface bears only tufts of rhizoids.

Turning to leaf form (already touched upon in *Fissidens*), we find great variety within the British flora, but for some of the most bizarre examples we have to turn to the mosses of other lands. It would be pointless to attempt a catalogue of the diverse leaf forms met with. A study of the illustrations in any good flora will indicate something of the range—from narrowly linear to broadly ovate or sub-orbicular; from leaves that are straight and almost parallel-sided to leaves that (as in some species of *Drepanocladus*) taper to very fine points and are curved almost into a semicircle (cf. Fig. 12F). Leaf shape, character of margin and other morphological features, not to mention finer structure, afford a range of taxonomic characters of great importance. Parallel development must be widespread, however, and phyletic conclusions would seldom rest secure on leaf morphology alone.

Presence or absence of a midrib has been widely accepted as an important generic (or even family) character. A midrib tends to be present and is often highly organised in most orders of 'acrocarpous' mosses—for example in Dicranales, Pottiales, Eubryales and others. It is much less well developed and is often lacking altogether in many Isobryales, Hookeriales and Hypnobryales. An excurrent, hyaline 'hair-point'—composed of dead, air-filled cells—is not uncommon. Among British mosses it is widespread in the genera *Tortula* (Fig. 12A), *Grimmia* and *Rhacomitrium*, occurring sporadically elsewhere. W. Watson[450] has drawn attention to this (together with some other characters of moss leaves) as a type of xerophytic adaptation (cf. Chapter 9).

The frail leaves of mosses are often strengthened by a border of specialised cells, narrow and thick-walled in contrast with those of the rest of the leaf. *Mnium punctatum* and other species of the same genus (Fig. 12G) are good British examples. The whole 'pattern' of leaf cell structure is often highly characteristic of a particular group; witness the prevalence of papillose upper leaf cells in Pottiales; the hexagonal-rhomboid leaf cells in *Bryum* and related genera; the exceedingly long, narrow leaf cells in *Drepanocladus*, and so on. Indeed, we have here a rich source of taxonomic characters; and leaf cell structure has always to be examined when one is determining moss species.

To conclude these remarks on leaf form, one may refer to *Rhizofabronia sphaerocarpa*, from West Africa, which is figured by Brotherus[50] and must surely show one of the most extraordinary

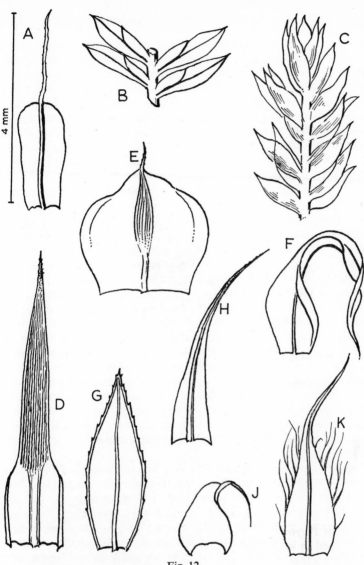

Fig. 12

Gametophyte of mosses: some types of leaf and shoot. A. *Tortula intermedia*, oblong leaf with truncate apex and nerve excurrent in long 'hairpoint'; B. *Fissidens bryoides*, part of shoot, with leaves strictly 2-ranked, each leaf consisting of boat-shaped clasping portion and expanded wing bisected by the nerve; C. *Isopterygium elegans*, part of ± flattened (complanate) shoot, with nerveless leaves not strictly in 2 ranks: D. Vegetative leaf, and E., perigonial leaf of *Polytrichum juniperinum*; F. *Drepanocladus revolvens*, strongly falcato-secund, nerved leaf; G. *Mnium hornum*, leaf with well-defined, toothed 'border'; H. *Dicranella heteromalla*, weakly falcate, nerved leaf (found in many Dicranales); J. Leaf of *Ctenidium molluscum*, strongly falcate, auriculate, nerveless; K. *Thuidium tamariscinum*, perichaetial leaf, with fimbriate margins. (All to same magnification.)

kinds of leaf to be found among mosses. The minute leaves could in a sense be described as pinnate, but the 'lateral members' are no more than extensively projecting individual cells.

It remains to consider the rhizoid system and then to say a few words concerning the special leaves borne on reproductive shoots. Many small, erect-growing mosses possess a well-developed rhizoid system which is in close contact with the substratum. Thus, when specimens of some of these small 'acrocarps' are collected for study it will be found that each little plant is provided with freely branched rhizoids which penetrate the soil to a depth at least equal to, and often exceeding, the height of the leafy shoot. This may be seen very well in some of the low-growing colonists of chalky soil, such as *Dicranella varia, Pottia davalliana,* various species of *Barbula* and others. Analogy with a root system is obvious, but in some ways a closer parallel is with root hairs. For the finer branches of the rhizoid system are very slender and quite colourless, with extremely delicate cell walls; also, they may become mis-shapen at their tips where in close contact with soil particles. The coarser strands are apparently both anchoring and conducting, and Goebel[148] explained their commonly oblique cross walls as an adaptation to facilitate faster conduction. These coarser strands are often brown, and sometimes their cell walls are papillose. Often there are perhaps half a dozen of these directed more or less vertically downwards and from them the obliquely placed, finer laterals arise which in turn bear the very delicate tertiary branches referred to above. A specimen of *Fissidens taxifolius,* which grew on soil in an Oxfordshire beechwood, proved revealing when carefully soaked out. The leafy shoots attained a height of 1 cm, the matted rhizoid system a depth of fully 3 cm.

Other mosses are remarkably deficient in rhizoids. This is true especially of many freely branched Hypnobryales where close contact with the substratum seems to be lost at an early age and the straggling older shoot systems are not anchored at all. Reference has already been made to the absence of rhizoids from mature *Sphagnum.* Elsewhere a dense covering of rhizoids clothes the stem for much of its length. This has been variously termed 'tomentum' and 'radicles' by systematic bryologists. Such rhizoids clothe the stems of *Polytrichum alpestre* in a dense whitish felt. They are well developed in some forms of *Dicranum scoparium*; and may be seen in many other mosses. The dense and almost continuous rhizoid covering of the stem in *Aulacomnium palustre* has often been cited as a good specific character. However, as mentioned by Sayre (in Grout's *Moss flora of North America,* vol. II[157]) and emphasised by Wallace,[438] a rather similar, though less dense, rhizoidal covering is also found on the stems of the arctic-alpine *Aulacomnium turgidum*.

In all these cases the 'tomentum' is a dense tangle of rhizoidal branches of varying calibre. The most slender may be only 3 μ in width and are sometimes swollen at their tips. Unlike subaerial protonema filaments they are not green. Rhizoid systems in which laterals come to lie along the same axis as main strands were detected in some species of *Polytrichum* more than a century ago. They have been likened to 'twisted string', and Goebel has commented on their manner of functioning like a wick in conduction. Wigglesworth,[464] much more recently, has demonstrated that they can also be important in vegetative reproduction (cf. Chapter 7). There can be little doubt that any moss stem bearing a dense rhizoidal covering is well equipped for external capillary conduction of water. It is also possible to visualise the significance of such rhizoids for water retention and absorption, for the main strands are direct outgrowths of superficial cells of the stem. Performing a like function, but much less widely distributed among mosses, are the paraphyllia. These clothe the main stems of *Hylocomium splendens, Thuidium tamariscinum* and some others. They are like minute, richly branched leaves of variable and irregular shape.

In many instances the leaves surrounding groups of reproductive organs are little different from vegetative leaves. In *Bryum, Funaria* and other genera the leaf size increases upwards on the stem, the largest leaves forming a group around the insertion of the sex organs. Those surrounding archegonial groups are known as perichaetial leaves; those investing antheridial clusters as perigonial leaves. In a single species the two may be widely different in form. Thus, in *Polytrichum* species the perigonial leaves are much the more strongly modified, and each consists of a very broad sheathing base and a short bristle point (Fig. 12E). In *P. juniperinum* they are red-brown or olive; in *P. piliferum* dull red. One sees the same marked change of form in the leaves that compose the bud-like 'male inflorescences' of autoecious species of *Bryum* and *Pohlia*; or again in *Philonotis* where the broadly expanded perigonial leaves give the antheridial receptacles a flower-like appearance and also provide a useful taxonomic character. The leaves that protect the antheridia in some species of *Fissidens* lack the characteristic 'wing' of that genus.

Perichaetial leaves, too, can be strikingly modified. Those investing the archegonia in many Hypnobryales are often nearly colourless, concave and long-acuminate; they must be efficient protective structures. They serve also to hold water—so essential to fertilisation in bryophytes—and this has been taken to be the explanation of the laciniate tips of the perichaetial leaves in *Diphyscium*, and the even more strikingly 'ciliate' perichaetial leaves of some species of

Thuidium (Fig. 12K). One need not emphasise that the chief work of the moss gametophyte is to bear sex organs; and it is not surprising therefore that many adaptations of leaf form have been evolved for their efficient protection and water supply. The form, arrangement and association with paraphyses of the sex organs themselves will be considered in Chapter 8.

In conclusion one must admit that, for all their diversity, the gametophytes of mosses do not compare in range of form with those of liverworts. Admittedly, they present many notably different kinds of habit, some of them very arresting. In this country we know the 'miniature trees' of *Climacium,* the 'moss balls' of *Leucobryum* and the festooning mats of *Neckera crispa,* to mention only three. Others are better seen elsewhere, and Martin[284] in his notes on the moss flora of New Zealand has referred graphically to forests where mosses are of real importance. He alludes to the 'yellowish-grey streamers of *Weymouthia*', and the 'golden tresses of pendent *Papillaria*'. Then again, there are the very large and the excessively minute among mosses; but if one measures diversity by the range of form exhibited by fundamental component structures the conclusion is inescapable. For leafy liverworts offer entire, two-, three- and four-lobed leaves; elaborately ciliate leaves such as those of *Trichocolea* and *Ptilidium*; complicate-bilobed leaves with elaborate sac formation as in *Frullania*; and mosses show none of these developments. True, in many mosses there is a complexity of internal organisation in the leaf (and in the stem) unmatched in any liverwort, but the underlying diversity is lacking. Parihar[320] has estimated that there are some 14,000 known species of Bryidae, as against 7,000 leafy liverworts. Yet a study of their gametophyte structure reveals them as a relatively circumscribed group.

The student who would enlarge his acquaintance with moss gametophyte structure should see, and examine microscopically, as much fresh material as possible; he should explore the resources of a herbarium, and this on a more than insular scale. He will also do well to consult some of the numerous monographic treatments that have appeared in recent years. Examples, to mention only a few, are the revisions of Japanese *Hypnum* by Ando,[11] of *Homalothecium* in western North America by Lawton,[249] of *Polytrichum* by Osada,[314] of *Calymperes* by Reese,[355] of the Hookeriaceae of the USA and Canada, and more recently still, of Mexico, by Welch.[456] Then he will see how skilled taxonomists handle available information about the different structural features considered above. Other aspects of gametophyte structure are taken up in later chapters.

6

THE SPOROPHYTE OF MOSSES

The sporophytes of Musci (mosses) show the same fundamental parts as those of Hepaticae, but there is a difference of emphasis. The foot is less often bulbous or anchor-shaped, more usually dagger-like in form. The seta is a stronger and much longer-lived structure than in liverworts. Lengthening early, it contrasts with the seta of liverworts which lengthens only after the spores are mature, sometimes (e.g. *Pellia*) actually after they have germinated in the capsule. Also, there is greater tissue differentiation within the seta of a moss. Though slender, it is tough, and in some cases (e.g. *Pohlia nutans*) may persist long after the capsule has matured and perished. The most notable divergence, however, occurs in the capsule itself, which in a typical moss is an important organ of photosynthesis for much of its life. Haberlandt stated that the chlorophyll content of one capsule of *Funaria hygrometrica* was roughly equal to that of fourteen leaves (of the gametophyte). Stomata are normally present. Although sporogenous tissue (in all but *Sphagnum*) is derived from internal layers (endothecium), there is a well-defined columella. Hence the capsule of a typical moss is nearer to that of *Anthoceros* than to any other liverwort, although there are very important differences, such as the absence of a basal meristem and the entirely different mode of dehiscence. Moreover, elaters are unknown in the Musci. Thus, the moss sporophyte, in all normal examples, is only partially dependent on the gametophyte; and it is more highly organised internally than any liverwort with the possible exception of *Anthoceros*.

Even so, it is difficult to make generalisations that hold good for all mosses; and we may mention now some of the important exceptions, where the sporophyte structure differs from that outlined above. As might be expected, two of the most notable are *Sphagnum* and *Andreaea,* each the genus which typifies a whole order of mosses, and each rather far removed from all others. On wet moors and bogs during June one may find some species of *Sphagnum* rather commonly fruiting, and the pale greenish, nearly globose capsules can be detected amongst the leaves of the terminal cluster of shoots. By mid-July, or sometimes rather earlier, each capsule will be raised on a short stalk, or pseudopodium, of gametophyte origin, for there is no true seta here. The capsule itself will have become dark brown, and as it dries in the sunshine it shrinks until the air that it contains is held under considerable pressure. With mounting pressure of air within there comes a moment when the small, convex lid is blown off and the spores escape in a cloud. The air, hitherto confined, is suddenly released, and Ingold[203] refers to it as the 'air gun' mechanism of spore discharge. Lack of seta and mode of dehiscence are but two of the peculiarities in the sporophyte of *Sphagnum*. Two others are the dome-shaped spore sac and the development of sporogenous tissue from a superficial layer—the amphithecium. This last is a character in common with *Anthoceros* and a point of difference from all other mosses.

Andreaea agrees with ordinary mosses in having internal (endothecial) origin of spore-producing tissue; but on all other counts it is just as unusual as *Sphagnum*. Thus, the spore sac extends like a dome over the central columella instead of surrounding it in the form of a cylindrical sheath, as in other mosses; and a gametophyte pseudopodium again functions in place of a true seta. The most striking peculiarity of the sporophyte of *Andreaea,* however, lies in the capsule itself. This is minute and ovoid, tapering a little at base and apex. Longitudinally it is marked by four lines of weaker cells and as the capsule ripens its wall splits along these four lines, whilst remaining intact above and below. The four valves that result (Fig. 14E) are very sensitive to changes in moisture. If one examines mountain rock faces, where *Andreaea rupestris* and *A. rothii* are often fertile, on a dry day, the capsule valves can be seen widely gaping; but, once thoroughly wetted, they close up and the capsule resumes the ovoid form it had before dehiscence. Measuring only *c.* 0·5mm in length, the capsule of *Andreaea* must be among the smallest known, and its chlorophyll content appears meagre. There can be no doubt that the placing of *Sphagnum* and *Andreaea* respectively in two quite separate subclasses of Musci is amply borne out by this rapid review of the salient features of the sporophyte of each.

By contrast, *Funaria hygrometrica* can be taken as typifying a host of ordinary genera of the Bryidae (Eubrya of Smith). For in this large group the sporophytes of different genera, or even families, differ in only minor points. All agree in that sporogenous tissue is formed from the innermost layers of the early embryo; in all the spore sac is cylindrical, not dome-shaped; and all show the expected differentiation into foot, seta and capsule. There is no need to repeat in detail here the well-known sequence of events in *Funaria,* for it is to be found in almost every general textbook, and further details of its sporophyte are given in the New Biology series of 'Famous Plants'.[445]

To trace the stages from tiny ellipsoidal embryo encased in archegonial venter, through the spindle-shaped middle period when the calyptra has swollen in advance of the capsule, to the functional stage of the pear-shaped green capsule, makes a fascinating study. It can be carried on through the stages of waning photosynthetic activity, as the capsule turns first yellow, then orange and finally dark brown, and the lid is forcibly removed by the swelling annular cells and the spores are shed. This last process must often be very gradual, controlled as it is by the hygroscopic movements of the outer ring of peristome teeth. Two unusual features here are that both sets of sixteen teeth are on the same radii and that the outer teeth are united terminally in a little disc of tissue. A remarkable fact about *Funaria hygrometrica,* and one which is very useful to the teaching botanist, is its capacity to show a wide assortment of different sporophyte stages within a single colony. Greene and Greene[152] have drawn up a scheme for comparing different mosses with respect to the 'time-table' of events in their life cycles. It would be interesting to see what kind of analysis emerged for *Funaria,* for it surely shows exceptional latitude. We must pass on now, however, to give a comparative account of the sporophytes of the Bryidae. It is convenient to distinguish between (1) early stages, (2) the functional green stage (of capsule) and (3) the stage of dehiscence and spore discharge.

(1) *Early stages.* According to Campbell[64] the majority of Bryidae which were examined embryologically showed close agreement in the early stages of the sporophyte. Polarity is acquired at the outset and the young sporophyte (whatever its ultimate form) elongates rapidly through the activity of an apical cell. Normally a second apical cell functions at the lower extremity of the sporophyte, and the whole structure changes from ovoid to ellipsoid, and finally to a narrowly cylindrical form, tapering at each extremity. The foot penetrates deeply into gametophyte tissue and, as in *Funaria,* the external differentiation distally into seta and capsule is delayed

(Fig. 13L). Transverse sections through the young embryo show that periclinal divisions early mark off the central endothecium from the peripheral amphithecium; and thereafter the forerunners of spore mother cells (archesporium) become recognisable within the endothecium, but not until the whole structure is some twenty cells in thickness, and both central columella and external wall are well differentiated. It would appear from Campbell's figures that this degree of differentiation is not reached until the capsule starts to swell. The seta of course begins to lengthen fairly early in mosses, and in species like *Ceratodon purpureus* and *Pohlia nutans,* which over-winter with young sporophytes, a colour contrast can be observed before the capsule has begun to swell. At this time the reddish-purple seta contrasts with the pale green terminal capsular region. Subsequent changes in the capsule are more marked in *Pohlia* than in *Ceratodon* for in the former it is finally pyriform and pendulous, whereas the capsule of *Ceratodon* (or that of *Tortula*) remains cylindrical and erect. Where the ultimate shape of the capsule is sub-spherical, as in *Bartramia,* more rapid and extensive changes must take place at a certain critical stage of development.

Rather little is known of the mechanisms controlling the steps whereby the sporophyte goes forward from an undifferentiated early stage to one in which we see foot, seta and capsule all well-defined; but the researches of Bauer[21] and others are suggestive (cf. Chapter 9). Clearly, the speed of events must vary greatly. In early winter ephemerals such as *Phascum cuspidatum* all the steps from fertilisation to fully swollen green capsule must be accomplished in a few weeks; for it is one of those mosses which lack a prolonged green stage and pass quickly from capsule swelling to dehiscence and spore discharge. Elsewhere, fertilisation is often a summer event, and capsules do not ripen until the following spring (e.g. species of *Dicranella, Tortula, Bryum, Mnium,* etc.); whilst fertilisation and spore discharge are separated by some thirteen months in *Polytrichum.* In this connection more biological studies of selected species are required (cf. Bennet,[23] 1965, on *Tortula muralis*; Longton and Greene,[264] 1967, on *Polytrichum alpestre* in the Antarctic).

(2) *The functional, 'green capsule' stage.* 'Functional' here refers to a stage of maximum photosynthetic activity and that of course embraces but one aspect of the function of a moss sporophyte. Biologically it is of great importance in establishing some degree of independence for the sporophyte as regards sources of organic matter; and in the evolutionary context it is important because it shows us the moss sporogonium nearer to an independent green land plant than it is at any other time; and this may indicate its progress towards the Pteridophyta, or its derivation from that

group (cf. Christensen[83]) according to the trend of one's evolutionary thinking. Campbell[64] pointed out that this green sporophyte in mosses had a lower proportion of its tissues given over to spore production than had any other bryophyte. This to him indicated a derived condition, for his thought was in line with that of Bower,[47] who in 1908 had pointed to an 'origin of a land flora' by the sterilisation of ever-increasing amounts of potentially sporogenous tissue. Nowhere among bryophytes was so much tissue thus 'sterilised' as in the green capsule (together with the seta and foot) of mosses. When one turns from these views to those of Christensen[83] and others one sees that 'the wheel has turned full circle'; for the greener, the more highly organised and more nearly independent the sporophyte is, the more primitive it is thought to be. There is no common ground between the two points of view.

In range of capsule form at the green stage one finds considerable diversity; sufficient indeed to provide the taxonomist with some useful systematic characters. Important differences between different members of the Bryidae based on internal characters of the green sporophyte do not appear to have been recognised, although it is possible that they may exist. The foot, when examined in transverse and longitudinal sections, appears generally to be well differentiated into outer (haustorial), intermediate (unspecialised) and central (possibly conducting) tissues. The outer have dense contents and currently some light is being thrown on them through electronmicroscopic studies (Eymé & Suire[121]). Lorch[267] implied that only occasionally was the haustorial character of this peripheral layer really well marked. Eymé and Suire, however, leave no doubt that, in *Mnium cuspidatum* at least, these cells are the site of intense activity. The cells of the central region tend to be narrow and elongated. In longitudinal sections which I have seen (*Mnium, Bryum, Atrichum*) the foot was almost parallel-sided above, then tapering into the apex, and deeply penetrating. Concealed in life and in herbarium specimens, this biologically important organ has perhaps had less than its due of attention. Lorch[267] indicates that only occasionally is the haustorial character of the peripheral layer of cells really well marked. One has to turn to seta and capsule for well defined family or generic distinctions.

The seta varies from a minute structure in cleistocarpous genera such as *Ephemerum* and *Phascum* to a notably elongated organ specialised for support and conduction. In general, a seta over 5 cm long is exceptional. Even the massive seta of the tall *Polytrichum commune* seldom exceeds 5 cm by much; but I have measured a seta over 7 cm long in *Pohlia nutans,* and one exceeding 10 cm in *Drepanocladus fluitans*. In this last species sometimes only a very long seta will raise the capsule clear of the water in the deep pools that it inhabits. The seta of *Polytrichum commune* is almost 0·5mm in diameter; that of *Pohlia nutans*, in common with most other mosses, is only about 0·2mm in diameter.

This slender organ has been much used taxonomically because it can often give minor but clean-cut and constant differences between related genera or species, especially differences of colour or papillosity. Use is made of the former character in *Dicranella*, of the latter in *Brachythecium* and *Eurhynchium*, in separating pairs of closely related species. The papillae on the seta of *Brachythecium rutabulum* are large enough to be easily visible with a hand lens. The colour of the seta, characteristically red in *Dicranella varia*, and yellow in *D. heteromalla*, cannot always be relied upon. I have seen the seta of *Funaria hygrometrica* red, although it is normally greenish yellow, and when old or flooded, setae (like capsules) may become discoloured and blackish.

Most of the principal orders of mosses are marked by their own characteristic form of capsule. It is narrowly cylindrical and erect in such a genus as *Ditrichum* in the Dicranales, and in *Tortula* in the Pottiales. It is pyriform and pendulous in many species of *Bryum* and *Pohlia*, subglobose in *Bartramia* and *Philonotis*, and a short, curved structure in many genera of the Hypnobryales (cf. Fig. 13). Both shape and inclination (erect, inclined, horizontal or pendulous) are important characters taxonomically, but no systematic bryologist is satisfied with the green stage, since it lacks the highly significant details of peristome and spores. The real interest of the green stage is biological, and it is pertinent to enquire to what extent the various features seen in the green capsule of *Funaria* are present also in other mosses. A few examples may be cited.

One will usually find the principal regions represented, apophysis and theca (spore-producing region), and in the latter, 'water jacket' layers of wall, chlorophyllose tissue, air spaces, spore sac and columella. The main differences lie in the proportions of these (Fig. 13). Thus, in *Bryum capillare* the sporogenous zone occupies a small fraction of the total width of the transverse section, because the wall is massive, the air space (crossed by trabeculae) extensive, the columella wide. In the rotund capsule of *Bartramia* the wide air space and slender trabeculae are again notable. Also, in stout capsules such as these the green tissue is very well developed and its photosynthetic capacity probably high. Narrowly cylindrical capsules, such as those of *Amblystegium serpens* and *Tortula muralis*, have a much narrower air space between wall and spore sac, trabeculae are less developed and green tissue less extensive. There is a many-layered wall and a broad central columella, but the cells composing

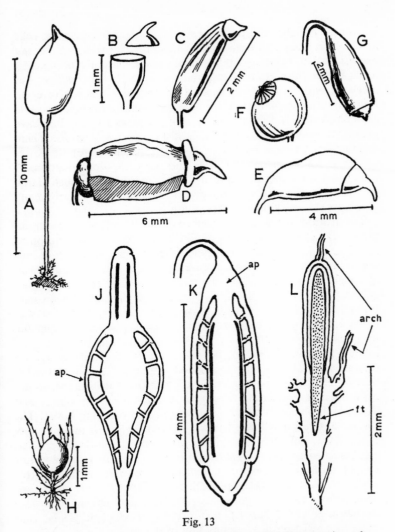

Fig. 13

Some types of moss capsule. A. *Buxbaumia aphylla*, disproportionately large, asymmetric capsule, long seta, rudimentary gametophyte. B. *Pottia truncata*, with operculum removed to show gymnostomy. C. *Ceratodon purpureus*, inclined, furrowed, slightly asymmetric capsule, with conical, shortly beaked operculum. D. *Polytrichum commune*, 4-angled, with discoid basal apophysis; E. *Eurhynchium striatum*, horizontal, with rostrate operculum; F. *Bartramia pomiformis*; G. *Bryum pendulum*—both with operculum removed; H. *Ephemerum serratum*, cleistocarpous capsule immersed in perichaetial leaves. J. and K., longitudinal sections of capsules; sporogenous tissue black, air spaces and trabeculae indicated diagrammatically; J. *Splachnum ampullaceum* with very large apophysis (ap) and K. *Mnium hornum*. L. Longitudinal section of a young sporophyte of a moss, semidiagrammatic. arch. archegonial neck; ft. foot.

both these regions tend to be smaller than the corresponding cells in *Bryum capillare*. In transverse sections of the capsule of *Tortula muralis* cut in February sporogenous tissue certainly occupied a much higher proportion of the whole than in *Bryum capillare*; but much depends too on just what stage has been reached in each. It is interesting to notice that the epidermal cells of the green capsule are comparatively thin-walled in *Amblystegium serpens,* but have greatly thickened and cutinised walls in *Tortula muralis.* There would seem to be a correlation with habitat here, for the latter species is found in notably drier and more exposed situations. I found that the columella cells of *Tortula muralis* held much starch at this stage, those of *Bryum capillare* little or none. Clearly many points would repay fuller investigation.

Two features of the green capsule which call for further discussion are the extent of the apophysis and the number and distribution of stomata. In many mosses the spore-bearing region passes over into the seta with little obvious development of neck (apophysis), although there will doubtless be more internal differentiation than is betrayed by the external form. Species of *Bryum* and *Pohlia* sometimes surpass *Funaria hygrometrica* in the development of a well defined neck region, but it is in the Splachnaceae that we find the most extensive apophysis of all. Members of this family are of interest for several reasons, not least because they mostly live on organic substrata, and this specialised habitat is linked with the big development of apophysis. The two commonest British species are *Splachnum ampullaceum,* which lives exclusively on dung, and *Tetraplodon mnioides* which grows mainly on decaying bones. The former has very much the larger apophysis, which is indeed many times the size of the theca (Fig. 13J). In certain non-British species, e.g. *Splachnum luteum, S. rubrum,* it is larger still and forms a yellow or red parasol-like growth unmatched elsewhere in mosses (for plate see Wettstein, R.,[461] p. 297). It appears that this specialised apophysis is not only conspicuous in size and colour; it is also attractive to dung flies through a secretion which it yields; thus these flies assist in the dissemination of the spores.

Paton and Pearce[326] have undertaken a broad survey of stomatal occurrence, structure and function in British mosses, and they point out that where the apophysis is large (as in *Splachnum*) the stomata tend also to be numerous. It is difficult, however, to explain the widely different numbers of stomata found in different species where the capsules are of normal size and proportions. Thus, over 200 stomata per capsule were found in some species of *Philonotis,* but much lower numbers than this are usual. In some cleistocarpous mosses, e.g. *Pleuridium* spp., only three or four stomata were found per capsule. Rather strangely, capsules that lacked stomata were found in widely divergent groups of mosses. Some were aquatics, like *Fontinalis antipyretica,* but others were terrestrial, for example all species of *Atrichum* and some of *Poly-trichum.* Clearly it is possible for some moss capsules to have a satisfactory functional life in the absence of the openings to the exterior which stomata provide. Furthermore, the stomata often become permanently closed with the advancing age of the capsule, and Paton and Pearce remark that even when the capsule is young and green these stomata lack sensitivity to such influences as changes in light intensity or carbon dioxide concentration. During this period they provide access to the internal atmosphere of the capsule and, we are told, 'close only when the water content of the capsule itself is greatly reduced, owing to extreme drought'. In some species the stomata are deeply sunk beneath the surface of the epidermis, but the incidence of this shows no clear correlation with dry habitats. Indeed, in *Orthotrichum* one finds 'superficial' and 'immersed' stomata in species of like ecology. Such a clear-cut difference has proved useful taxonomically. The 'ring-shaped' stoma of *Funaria,* with the two guard cells confluent, is exceptional.

(3) *The stage of dehiscence and spore discharge.* With the gradual ripening of the capsule we have access to its most valuable taxonomic character, the peristome. This surrounds the mouth of the capsule after the lid has been shed. The teeth composing the single peristome of some mosses (Aplolepideae) and the outer peristome of others (Diplolepideae) are composed of a kind of two-ply material furnished by the cell walls of adjacent cells. Ordinarily sixteen teeth spring from the diaphragm at the mouth of the capsule, their bases forming a rather close-fitting circle, their apices tapering to quite fine points. There is much variety of detail. One may cite the twisted cone of *Tortula subulata,* the sixteen cleft teeth of *Dicranella heteromalla* (Fig. 14A), the eight pairs of reflexed teeth in *Orthotrichum,* to mention only a few. Almost always, however, the main ring of teeth (the outer peristome when this is double) consists of relatively firm, golden brown, barred structures, the bars revealing the limits of the cells whose walls have persisted to form the tooth; and owing to their two-ply structure and the uneven deposition of their thickening these principal teeth are sensitive to changes in humidity. When a complete inner peristome is present it consists essentially of a pale tubular membrane which supports a surprisingly elaborate superstructure. For arising from this basal membrane, and alternating with the sixteen outer teeth, are sixteen cleft or fenestrate inner teeth, whilst on the same radii as the outer teeth are groups of thread-like 'cilia' (Fig. 14H). These cilia commonly occur

three or four together, and when perfectly developed have short transverse appendages. For really fine illustrations of peristome structure and development the student should consult Ruhland.[370]

In some genera, for instance *Bryum, Pohlia,* one finds certain species that have a complete double peristome, with appendiculate cilia, but in the same genus there are others in which it is to varying degrees imperfect. In these the cilia may be represented by mere stumps and appear to be vestigial. Indeed, there is evidence that here and there, in widely different circles of affinity, the peristome has undergone reduction; and this has sometimes been so drastic that the structure has been lost altogether. Formerly bryologists placed all such mosses together in a separate group, but nowadays it is universally recognised that a moss which has lost its peristome may be closely related to other species in which this structure has been retained. Thus, to mention two examples, *Pottia lanceolata* is peristomate, *P. truncata* (Fig. 13B) is gymnostomous; in *Funaria hygrometrica* the peristome is well developed (and double); in *F. fascicularis* it is rudimentary or lost.

In moss classification great stress has always been laid on the peristome, especially since the time of Philibert,[335] whose classical papers were recently abridged by Taylor.[426] Distinct types of peristome will be found often to characterise whole orders, e.g. Dicranales, Pottiales, Eubryales. Yet, like all taxonomic characters, it must not be used inflexibly. Thus in the genus *Encalypta* there is no reason to doubt that the different species compose a thoroughly natural group. Yet among the British species alone one finds in *Encalypta streptocarpa* a double peristome, in *E. rhabdocarpa* a single peristome and in *E. commutata* no peristome at all.

Some moss capsules have not only lost the peristome but are without a detachable operculum or lid. These are known as cleistocarpous species and the 'closed fruit' opens eventually in an irregular manner to release the spores. The minute *Ephemerum serratum* (Fig. 13H), which may sometimes be found on mud at the margins of ponds, is a good example. It is generally held that even cleistocarpous species may be closely related to others that have peristomate capsules.

Turning to the question of spore discharge, one can say that after the lid has been shed the rate at which spores leave the capsule is determined by the peristome. Goebel[148] reviewed the more important arrangements and distinguished between what he called primitive, intermediate and complex types. An example of his primitive condition is provided by certain species of *Barbula* and *Tortula,* in which Goebel held that the long, spirally twisted peristome teeth functioned as a 'hygroscopic lid'. It is true that the peristome of such a species as *Barbula unguiculata* (Fig. 14C) assumes very different appearances in wet and dry conditions, but Goebel's point is that active movements of the teeth themselves play little or no part in effecting the dispersal. An intermediate case is that of *Dicranella heteromalla,* with its single ring of sixteen forked, freely moving teeth (Fig. 14A). Here the spores accumulating under the mouth of the shrinking capsule become caught among the slowly moving peristome teeth, and are shed. Goebel's 'complex' type is illustrated by the double peristome of such genera as *Bryum, Mnium* and many others, including a vast range of Hypnobryales. In most of these there is rather more active participation by the teeth in spore dispersal than one can see in the 'bowing' movements of the *Funaria* peristome. *Brachythecium velutinum,* described in detail by Ingold,[203] is a good example. Here, in dry conditions, the inner peristome stands up as a pale central cone and the tips of the arched outer peristome teeth are inserted into the gaps between the various inner structures. By such a device spores can be actively flicked out. *Hypnum cupressiforme* (Fig. 14G) is similar. It is usual for the spores in a ripe capsule to be massed near the mouth, and gradual shrinkage of the organ doubtless promotes this. Once this position of the spores is attained, the peristome can bring about their gradual discharge. The whole subject has recently been re-examined on a broad front by Pais.[316]

Ingold[204] has thrown light on some general questions relating to the mechanism of gradual discharge. He points out that in *Mnium hornum* the spores are not sticky and therefore tend to fall naturally towards the mouth of the pendulous capsule. As some are released, others will fall into position. He tested a capsule of *Eurhynchium confertum* by alternately breathing upon it and allowing it to dry, in order to see how effective was discharge by peristome tooth movement alone. The capsule was laid on its side on a glass slide; the procedure was repeated 171 times and a total of 15,647 spores were discharged. Such observations demonstrate the efficiency of the peristome mechanism for securing a gradual liberation of the spores; and, as Ingold points out, an important feature of the whole peristome apparatus is its power of closing the capsule in moist conditions, thereby eliminating spore discharge at such times. Lazarenko[250] saw in the peristome a device for securing spore discharge *at the right time,* which would be neither in wet nor in very dry conditions (e.g. periods of sunshine and showers). He alluded to Patterson's observation that some epiphytic mosses had peristomes which opened in wet conditions. He also drew attention to some exceptional cases, e.g. *Heterophyllium haldanianum,* in which the peristome is so fragile that after ten to twenty hygroscopic movements the teeth are shattered and the capsule acts as if gymnostomous. There is plenty of room for further observation and experiment.

Ingold reports total spore contents of capsules in *Eurhynchium confertum,* as determined by haemacytometer techniques, between 280,000 and 700,000. The capsule in this species is not notably large and these figures seem astonishingly high. My own rough estimates made from time to time on species of *Bryum* and *Pohlia* have led me to total figures much lower than this. Doubtless there is some correlation with size of spore. In many Hypnobryales (to which *Eurhynchium* belongs) the spores are indeed very small, with diameters of 7–10 μ. In the genus *Bryum* there is great diversity, for whereas many species have spores quite as small as the above, those of others attain 20–30 μ, and the spores of *B. warneum* may reach 50 μ in diameter (Fig. 15M). Records of spore outputs in these different species would be of interest. Despite some latitude as regards size, the spherical spores of the Bryidae seem to be relatively uniform structurally. Scanning electron microscopy however is beginning to reveal unsuspected diversity of exospore ornamentation[275].

We have seen that it is partly on account of their unusual sporophytes that the genera *Sphagnum* and *Andreaea* are made respectively the basis for separate subclasses, Sphagnidae and Andreaeidae. Rather surprisingly, there exist certain groups in the Bryidae in which the capsule shows important departures from the kind of structure described in the preceding pages. The most notable examples are *Archidium* and *Tetraphis.* Also, Polytrichidae and Buxbaumiidae display unusual types of sporophyte.

Turning to the first of these, we find in *Archidium* a most unusual moss. The gametophyte is ordinary enough, being much like that of numerous genera in the order Dicranales. The capsule, however, is not merely cleistocarpous; it is without a columella and produces only four to twenty-eight exceedingly large spores (200 μ in diameter). The foot is nearly spherical and there is no seta. Many authors have seen it as an archaic type. Even if reduced, it may well represent a very ancient line of descent to simplicity.

Tetraphis pellucida is interesting for several reasons (cf. Chapter 7), but not least for its capsules. They are erect, cylindrical structures and bear a superficial resemblance to those of such genera as *Tortula* or *Barbula,* but a lens reveals that the peristome is composed here of only four large teeth of solid construction (Fig. 14D), more akin to those of *Polytrichum* and its allies. They are, however, held erect in the ripe capsule and there is no trace of an epiphragm. They would not appear able to effect gradual dispersal of the spores. With so aberrant a peristome the order Tetraphidales seems mis-

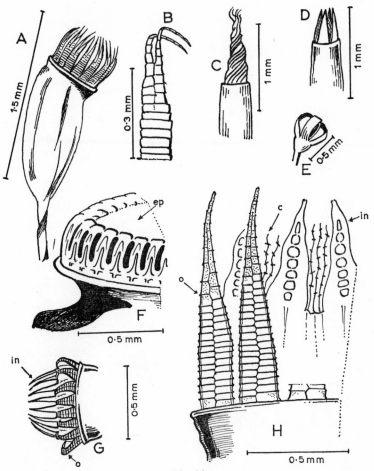

Fig. 14

Structure of peristome in moss capsules. A. *Dicranella heteromalla*, single ring of 16 barred teeth; B. single forked tooth from same; C. *Barbula unguiculata*, single ring of spirally twisted, almost filiform teeth. D. *Tetraphis pellucida*, peristome of 4 erect, 'solid type' teeth. E. *Andreaea rupestris*, no peristome but capsule opening by 4 valves. F. *Polytrichum commune*, part of mouth of capsule (operculum removed) showing epiphragm (ep) and a few of the slits through which spores escape. G. *Hypnum cupressiforme*, double peristome, with 16 barred outer teeth (o) and more delicate inner teeth (in). H. *Bryum capillare*, part of capsule mouth showing two complete outer teeth (o) and part of the complex inner cone which consists of fenestrate inner teeth (in) and appendiculate 'cilia' (c). A, C, D, E and G are all shown as they appear when dry.

placed among the orders of 'normal' Bryidae, in the system of
Reimers, but that author suggests that *Tetraphis* (with four species)
and the closely related *Tetrodontium* (one species) are less isolated
than at first sight appears (cf., however, p. 166).

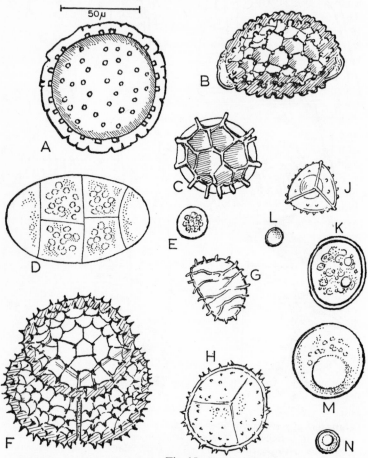

Fig. 15

Spores of some liverworts (A–H) and mosses (J–M) to show diversity of size,
form and exospore ornamentation. A. *Naiadita*. B. *Riccia glauca*. C.
Targionia hypophylla. D. *Pellia epiphylla*. E. *Lophocolea heterophylla*.
F. *Sphaerocarpos michelii* (tetrad). G. *Fossombronia pusilla*. H. *Anthoceros
husnotii*. J. *Sphagnum* sp. K. *Andreaea rothii*. L. *Polytrichum commune*.
M. *Bryum warneum*. N. *Bryum capillare*. Note in D precocious germination,
in F a persistent tetrad. In H and J note conspicuous triradiate mark.
D, E and K are markedly chlorophyllose, M and N moderately so. J, M
and N show conspicuous oil drop. All are to same scale.

We have seen in the last chapter that the Polytrichidae offered some unique features of the gametophyte. This is no less true of the sporophyte. For here, in place of a radial division of peristome into sixteen barred units (with or without the addition of inner structures) one sees thirty-two or sixty-four teeth of an entirely different type. They are peculiarly solid structures, each tooth being composed of fibre-like cells several layers in thickness. Moreover, these teeth are joined at their tips to a pale membrane, the epiphragm, which is stretched like the tympanum of a drum across the capsule mouth after the operculum has been shed. The spores are thus dispersed by a censer mechanism through the minute holes between successive teeth (Fig. 14F). With reason, it has been likened to a poppy capsule. *Polytrichum, Atrichum* and *Oligotrichum* all show this structure and the white epiphragm is a conspicuous feature of old capsules in many of the species; especially so perhaps in *Polytrichum aloides,* against the dark background of the shaded recesses where it grows.

Finally, in *Buxbaumia* (Fig. 13A) and the allied *Diphyscium* we find a capsule which immediately arrests attention on account of its large size in relation to the small gametophyte, and its very obvious asymmetry. The peristome is also unusual. In *Diphyscium* the outer peristome is defective or absent and the inner takes the form of a delicate, whitish 'pleated' cone, not closely matched elsewhere. *Buxbaumia* has the same type of inner peristome, but the outer consists, according to the species, of one or several rings of thread-like teeth. Several concentric rings of cells are involved in the formation of this outer peristome and it is evident that the teeth here cannot be homologous with those of ordinary peristomes. Dixon[107] alluded to these plants as representing most probably the survival of an ancient pattern of moss structure, and to supporters of Bower's antithetic theory of bryophyte evolution this would certainly be a reasonable view. Goebel's views on these strange mosses have been cited at an earlier point (cf. p. 68). There is much here to suggest a primitive survival. At all events, it would be difficult to present a convincing argument for the derivation of such a strange peristome from the typical structure found, with minor variants, throughout the great body of the Bryidae. We must not forget that the latter comprise more than 90% of all known mosses. It is surely the aberrant, so few and so unlike the rest, that represent the survival of other, less successful evolutionary lines. This is so in other great taxonomic groups, in flowering plants for example, and again in birds and mammals.

7

ASEXUAL REPRODUCTION

Asexual reproduction in bryophytes can be achieved in three ways: by growth and branching, followed by the death and decay of older parts; by the separation of whole organs and the regeneration of new plants from them; and thirdly, by means of specialised units of propagation termed gemmae.

The first method is more widespread than is generally realised. It is claimed, for example, that the plate-like protonema of *Sphagnum* bears initially but one gametophore. If this is so, then the dense tuft or cushion so characteristic of mature plants must have arisen by repeated branching and the subsequent death of older branches. A similar outcome can be seen in liverworts, for example in *Riccia*, where not only the rosette form of the mature gametophyte but also the vegetative spread of the plant may be achieved in this way. It is often impossible, however, to decide how far the mature condition of a moss results from the process just outlined and how far it is caused by numerous buds springing from the original protonema. In any event this is a process calculated more to increase the bulk of the plant than to allow of its spread to entirely new ground. It is scarcely vegetative propagation in the full sense.

The separation of whole organs, on the other hand, can be a most effective means of spreading a plant. The best known examples among mosses are whole shoots or shoot tips, and two common heathland mosses will serve as illustrations. In the first, *Pohlia nutans,* one can often find numbers of the somewhat catkin-like deciduous branches scattered over the surface of the 'short turf' which this plant forms. They appear light against the dark green of the parent moss and the leaves which they bear are small and closely appressed. Even commoner is the sight of the deciduous, light-coloured shoot tips on the deep green patches of *Campylopus piriformis,* a colonist of bare peat and decaying stumps. In this species and the closely related *Campylopus flexuosus* rhizoids are produced freely at the leaf base; these no doubt facilitate regeneration from the deciduous shoot tips.

It is likely that many species which are widely distributed despite infrequent production of sporophytes are able to spread by means of deciduous shoots or shoot fragments in this way; but we have little precise information. Species that spring to mind are *Pleurozium schreberi* and *Pseudoscleropodium purum.* The former is a very common moss on heather moor, the latter the most widespread species generally on chalk grassland. Yet in neither are sporophytes common and in *Pleurozium* they are decidedly rare. Tamm[421] in his detailed study of *Hylocomium splendens* was led to suspect some such means of vegetative spread in that species and was puzzled too by the rarity of young stages in the field. *Myurium hebridarum,* a locally plentiful moss in the Outer Hebrides (and found in a few scattered stations on the Inner Hebrides and adjacent mainland), undoubtedly must spread by deciduous branches, for the sporophyte has never been found in Britain.

Degenkolbe,[99] in a comprehensive survey of the organs of vegetative propagation in leafy liverworts, recognised the importance of deciduous branches (Brutäste) and leaves in this connection. He concluded that deciduous branches were an important means of spread for many genera of epiphyllous liverworts in the tropics, all of them members of the family Lejeuneaceae. As examples of deciduous leaves effecting propagation he cited exotic species of *Plagiochila, Bazzania* and *Drepanolejeunea,* also two British liverworts, *Mylia cuneifolia* and *Frullania fragilifolia.* Degenkolbe described how, in *Frullania fragilifolia,* the large, flat antical lobe of the leaf would be shed as a propagule, leaving rows of the small postical lobes intact along the stems. These also could be induced to give rise to new plants. He grew them on sterile filter paper in Knop's solution in a Petri dish and obtained a cell mass at the cut surface after a few days. From this cell mass a young plant arose. According to him, the well-known deciduous perianth of *Gymnocolea inflata* is unique. Anybody familiar with this common plant of wet heaths will know how plentiful such perianths are. Acting as propagules, they must be an important factor making for the abundance of the species.

Cavers,[71] in a useful summary of asexual reproduction and regeneration in Hepaticae, drew attention to a rather different process, intermediate between the two described above. It is well seen in thallose Marchantiales and consists of the production of 'adventive branches' (often from the underside of the midrib) and the subsequent breaking free of these to form new plants. Cavers noted how a whole pond surface could become covered by *Riccia fluitans* in this way, and how, in some xerophytic species of the same genus, tuberous branchlets thus released enabled the plants to survive periods of drought. He noted a similar process in *Targionia,* stating that 'the ventral shoots become detached at the base and form new plants'. This is comparable with the spread of a stoloniferous or rhizomatous vascular plant; no more than that, for there is little prospect of the new thallus being carried far by wind or water in the way that microscopic deciduous propagules could be.

It will now be convenient to consider some cases among mosses where the organ shed is a modified branch of bud-like form. This type of organ, which occurs in some species of *Bryum* and many species of *Pohlia,* is strictly a deciduous branchlet, although it is variously termed a 'gemma' or a 'bulbil' in taxonomic works. In *Bryum bicolor* var. *gracilentum* there is no mistaking the shoot-like character of the propagules, and the stem with its leaf rudiments may be lengthened to varying degrees. So, too, the 'bulbils' of *Pohlia bulbifera* and *P. rothii* are clear foreshortened swollen axes with rudimentary foliar appendages. In the curious, sometimes 'glove-shaped' propagules of *Pohlia annotina* the distinction between axis and leaf-like appendages is less clear. In each of the above examples the precise form of the bulbils constitutes an important taxonomic character. Five or more bulbils per leaf axil are common in *P. annotina* and they must be a principal means of propagation; yet in none of these species is the sporophyte unknown, although admittedly it is relatively rare.

Although Necker described experiments on regeneration as long ago as 1774 in his *Physiologia Muscorum,* it was not until the latter part of the nineteenth century that the subject attracted widespread attention, culminating in the reports of Vöchting,[435] Schostako-witch[379] and Cavers.[71] Kreh[241] went on to explore the subject in a really comprehensive manner. In a long paper devoted to regeneration in liverworts he showed that almost every part of the plant, with the exception of antheridia, could be induced to undergo regeneration. He had success with isolated perianths, with archegonia, even with the ventral scales of Marchantiaceae and isolated rhizoid cells. In short, he was able to show how almost any organ or part of a liverwort plant which became severed from the parent body would be able, given suitable conditions, to regenerate in time a whole plant. The implications of this for vegetative spread are obvious. In fact Vöchting had already stated, for *Marchantia,* that every living cell is capable of regenerating the entire plant, and Goebel's statement that 'almost every living cell of a moss can grow out into protonema' carries a similar implication for mosses. Surely, no accident which can befall a clump of growing moss or liverwort (and result in fragmentation) will be without the possibility of beneficial consequences in the shape of subsequent spread. In more recent years Wettstein[457-8] and others have drawn on the truth of Goebel's dictum to encourage cut pieces of seta to form protonema and hence diploid gametophytes. Fulford[131-2] too has shown how the application of very dilute growth-promoting substances can greatly facilitate regeneration in hepatics. Still more recently the subject of regeneration in lower plants has been reviewed by Stange,[408] but the morphogenetic questions she raises are more appropriately considered in Chapter 9. Suffice it here to say that the exceptional regenerative powers of bryophytes cannot fail to be of use in propagation.

As already indicated, the term 'gemma' is used in a rather wide sense by some authors; but more properly it is restricted to a propagative organ of definite form and quite unlike the parent plant from which it springs. It characteristically originates from a single cell, but may be unicellular, bicellular or multicellular at the time of its release. It is not organised into a short axis and rudimentary leaves. If one accepts this last criterion one excludes the bulbils of most of the *Pohlia* species described above, but one retains a diversity of reproductive units. They include the large discoid gemmae of *Marchantia* and *Lunularia* which are so efficient that Goebel was able to say, with slight exaggeration, that they 'over-run every pot in cultivation'. They include a range of globose, discoid and plate-like structures of intermediate size, some being found in mosses, others in certain groups of liverworts; and a further group of units (found among mosses) whose filamentous form suggests derivation from protonema. Finally there is the great mass of leafy liverwort genera producing one- or two-celled gemmae on leaf points or modified stem apices. The masses of axillary structures formed in the moss *Isopterygium elegans* are probably better regarded as highly modified deciduous branchlets. Each consists of a length of tenuous stem bearing rudimentary leaves.

For the student wishing to study a small but reasonably representative range of these structures one might suggest the following: among liverworts, (1) the large discoid gemma of either *Lunularia* or *Marchantia,* (2) the small plate-like gemmae of *Metzgeria fruticulosa,* (3) *Blasia pusilla*—with two kinds of propagules, and (4) *Lophozia ventricosa* with its microscopic one- or two-celled gemmae; among mosses, (1) the multicellular gemmae borne on the characteristic receptacles of either *Tetraphis pellucida* or *Aulacomnium androgynum,* (2) microscopic filamentous gemmae of either *Ulota phyllantha,* or *Zygodon viridissimus,* and (3) ovoid or subspherical 'rhizoid gemmae' in either *Leptobryum pyriforme* or *Bryum rubens.* Most of these examples should prove readily obtainable. A little may now be said about each.

The biconvex gemmae of the liverwort *Lunularia cruciata* are nearly 0·5mm in diameter and are just visible to the naked eye. They are numerous in each receptacle (Fig. 16A) and gemmae of many different ages can be seen at any one time. When ripe they become readily detached from their short stalks, a process facilitated by the club-shaped mucilage hairs that occur on the receptacular surface. Close behind the apex of the thallus the gemmiferous receptacle becomes exposed by the raising of a flap of tissue on the upper surface of the thallus. Vertical longitudinal sections show that at first the receptacle is a deep hollow, almost concealed between the nearly vertical flap and the upturned apical part of the thallus. Later the receptacle is displaced backwards, the flap becomes a slope and the receptacular surface itself may be almost vertical. It is easy to see how this adjustment could assist in the washing away of the ripe gemmae. These structures, bright green and rich in starch and oil, have been described and figured again and again. They are remarkably similar in the two genera. It is worth examining the soil in the neighbourhood of gemmiferous plants in search of early stages in the germination of the gemmae. One may be found that has widened considerably by growth from both its laterally placed growing points and put forth a tuft of long slender rhizoids from the central 'cushion' (Fig. 16D).

Metzgeria fruticulosa is generally accorded specific rank nowadays, although it used to be classed as a variety of *M. furcata.* It is abundantly gemmiferous (Fig. 16E) and the branches which bear the gemmae are marked by narrow form and nearly erect growth. Gemma production in this genus has been well described by Evans.[118] *M. fruticulosa* is an abundant epiphyte in western Britain, and will repay careful study. Many stages in the origin of gemmae will be seen without difficulty. It will be noticed that each of the flat, roughly circular plate-like gemmae originates from a single superficial cell. Fully grown, the gemma (Fig. 16F) is about 0·2mm in diameter. It does not exceed one cell layer in thickness and there is no marked differentiation of cells. However, at the pole remote from the attachment a distinct apical cell can usually be detected. Some marginal cells may grow out to form rhizoids even before the gemma is shed. This type of gemma resembles those described by Degenkolbe[99] for various members of the family Lejeuneaceae.

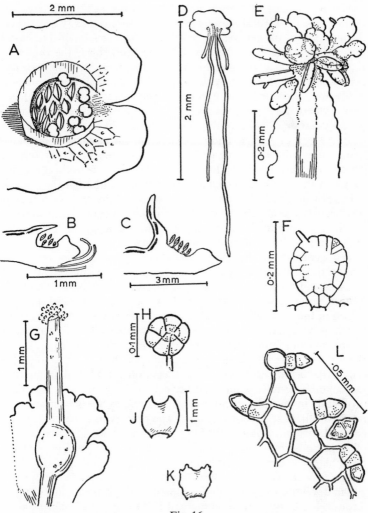

Fig. 16

Asexual reproduction in liverworts. A–D. *Lunularia cruciata*. A. Gemma receptacle, from above. B and C. Early and later stages, gemma receptacle in longitudinal section. D. Gemma, with rhizoids, in process of development into new plant. E–F. *Metzgeria fruticulosa*. E. Cluster of gemmae forming at extremity of thallus lobe. F. Single gemma, with indications of apical cell and rhizoid initial. G–H. *Blasia pusilla*. G. Portion of thallus with flask-shaped gemma receptacle. H. single gemma. J–L. *Lophozia ventricosa*. J. Normal, and K. gemmiferous leaf. L. Gemmae arising from marginal cells of K, and one free gemma.

The propagules of *Blasia pusilla* have long been of interest to bryologists, for several reasons. First, this species has two perfectly distinct kinds of gemma; second, the production of one kind makes the plants extremely conspicuous; and third, the origin of the curious tubular or funnel-like organs responsible (Fig. 16G) has been something of a mystery. Unfortunately, *Blasia* is not a common liverwort, but once found it is worth cultivating in the cool green-house, where it will grow successfully for a number of years. Cavers[71] gave a good description of the mode of origin of these two perfectly distinct kinds of gemma, and indicated how the swelling of mucilage hairs brings the spherical gemmae to the mouth of the funnel. These gemmae, on their long slender stalks, are akin to those found in the moss *Aulacomnium androgynum,* although their shape (Fig. 16H) is different. Those of the second type are scale-like. It has been suggested that the tubular receptacles of *Blasia* may be modified archegonial receptacles, on the evidence of occasional archegonia which have been found in them.

The gemmae of the leafy liverwort, *Lophozia ventricosa,* are among the simplest known, for they are one- to two-celled structures, usually the latter. They are formed in great numbers, either at each point of some of the bifid leaves or on the modified apical region of the axis. Some would restrict the term gemma to this type, which is widespread among leafy liverworts. When a leaf point becomes gemmiferous certain of the marginal cells begin to protrude, the walls concerned becoming clearly thinner than those of adjacent cells. After a time a transverse wall cuts off a stalk cell from what Degenkolbe terms a gemma mother cell. This mother cell may then proliferate in several directions (Fig. 16L), and as mature gemmae are released others arise beneath them. The process has been likened to conidial formation in certain fungi; and in the longitudinal section of a gemmiferous 'head' of *Calypogeia* figured by Cavers[71] the resemblance is certainly striking.

In *Lophozia ventricosa* the gemmae usually become two-celled late in development. Often dense content will distinguish the gemmiferous cells. At times there is marked periodicity in gemma production, as Buch[55] pointed out in *Scapania.* Sometimes the leaves bearing gemmae in *Lophozia ventricosa* are little altered in form, except that the two points bear clusters of these microscopic propagules; at other times they become erose and highly irregular in outline (Fig. 16K). Indeed, gemma production here seems to lack the precision that one associates with it where the gemmae are more complex.

Turning now to mosses, we may begin with *Tetraphis pellucida,* which is often abundant on decaying stumps. It grows as a 'short turf' (cf. p. 139) and a group of much enlarged leaves normally (but not invariably) marks the terminal gemma receptacle (Fig. 17A). When one dissects this in a drop of water gemmae of varied age escape, perhaps twenty to thirty mature ones and many that are much younger. It is odd that these obviously immature stages should become detached so readily. Moreover, the same phenomenon is seen in *Aulacomnium androgynum,* where a parallel type of gemma occurs.

In *Tetraphis,* at quite an early stage, most of the thirty-five to forty cells of the mature, roughly heart-shaped gemma are present. The difference is in their smaller size and much less mature state. In very young stages (diameter *c.* 25 μ) the cells are about 10 μ wide, with very delicate walls and minute plastids (Fig. 17B). The mature gemma (diameter *c.* 150 μ) has cells about 20 μ wide (Fig. 17C), with firm walls and full-sized chloroplasts, although at first the apical cell, and later those specialised cells whence the 'germ-tubes' eventually come, will appear different. Part of the central region becomes two cells thick and the five or six cells composing the stalk lengthen abruptly at a late stage.

Enough has been said to indicate that in the well-known gemma cups of *Tetraphis* there are many features calling for close study. There is still room here for further observation, and even more for experimental work. In the neighbourhood of gemmiferous plants of *Tetraphis* one may find some of the young stages of the thalloid structure into which the gemma develops on germination. As was shown by Correns,[90] this may achieve varying form and ramification before leafy shoots eventually arise.

The stalks, or pseudopodia of *Aulacomnium androgynum,* with their terminal cluster of gemmae, appear like so many greenish pins (Fig. 17E) arising from the leafy shoots; this makes the species unique among British mosses, and it is very easily recognised. It grows on decaying stumps. In most essential features the gemmae resemble those of *Tetraphis,* although they differ in shape. Thus, they are borne on multicellular, delicately filamentous stalks, and the central part of the gemma (Fig. 17F) becomes two cells thick at maturity. The pattern of their further development, however, is quite different and a new leafy shoot arises relatively early. Ruhland[370] provides detailed accounts of both these examples, as well as superb illustrations of a wide range of moss gemmae.

Gemmae which consist of short filaments, composed of rather large, often relatively thick-walled cells, are not uncommon among epiphytic mosses. A striking example is *Ulota phyllantha,* where the gemmiferous surface is the projecting tip of the leaf (Fig. 17K,L). This moss is not confined to trees, but will also grow on rocks, often those near the sea and sometimes not far above high-water mark. While the leaf is still young, some of the cells in the shortly excurrent nerve grow out to form the filamentous, brownish-green gemmae. Thus, leaf tips may appear fully mature and may bear the characteristic clusters of gemmae when the cells in the leaf base are scarcely beyond the embryonic stage. Release is by the rupture of the basal

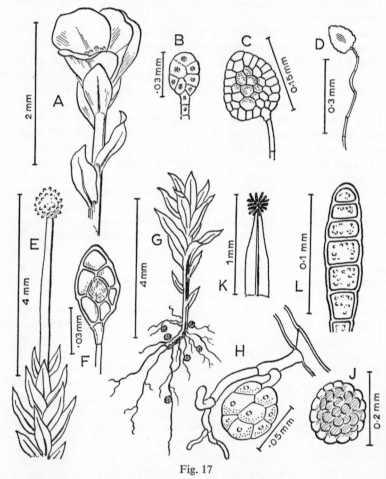

Fig. 17

Asexual reproduction in mosses. A–D. *Tetraphis pellucida*. A. Gemma 'cup' formed of 3 enlarged leaves. B. young, and C. mature gemma (omitting lower part of stalk). D. Gemma with complete stalk. E.–F. *Aulacomnium androgynum*. E. Stalk (pseudopodium) bearing gemmiferous head. F. Single gemma. G–J. *Bryum rubens*. K–L. *Ulota phyllantha*. K. Leaf, still young, with apical tuft of mature gemmae. L. Single gemma.

cell in the short filament and great numbers of broken cells can be seen projecting from a leaf apex which has shed its gemmae. Gemmae of this kind are widespread, especially among epiphytes. The closest parallel to *Ulota phyllantha* lies perhaps in the leaf tip clusters found in the big tropical genus *Calymperes*. In *Zygodon* they are club-shaped and in this genus and *Orthotrichum* the gemmae arise in variable numbers on the adaxial surface of the leaf. Filamentous gemmae of quite another type arise sometimes from the basal parts of leaves in *Encalypta streptocarpa,* and different again are the clusters which spring from either stem or leaf surface in several species of *Plagiothecium*.

The European mosses which bear gemmae on the rhizoid system have been the subject of a recent paper by Whitehouse,[463] who applies the apt name 'tuber' to this class of structure. They occur in a surprising range of species, referable to numerous genera. Almost all grow on soil and usually the tubers are subterranean. Our example is *Bryum rubens* (Fig. 17G), which is fully described by Crundwell and Nyholm[95] (in their account of the European species of the *Bryum erythrocarpum* complex). It is commonest on basic soil and forms nearly spherical, bright red gemmae both within the soil and in leaf axils just above soil level. Surface cells of the gemma are convex; which together with the colour, reminds one of the drupels of a raspberry. When ripe they are about 0·25 mm in diameter and readily detached.* A careful examination of the rhizoid system, however, will reveal others that are still attached, whilst further gemmae will be seen in various stages of development. In the early stages they are grey in colour, and the cells thin-walled with dense granular contents (Fig. 17H). Even then each consists of a solid 'knot' of cells. Each gemma appears to be the swelling of a short tertiary branch of the rhizoid system.

Leptobryum pyriforme has ovoid gemmae on its rhizoid system. They are of the same general character as those described above. Greenhouse pots are the favourite habitat of this moss. Whitehouse drew attention to the characteristic form of the tubers in different examples. In *Dicranella varia* they are narrow, few-celled and comma-shaped. Although known from certain species of *Dicranella, Ditrichum, Barbula, Tortula* and *Fissidens,* and from many species of *Bryum,* tubers have not yet been convincingly demonstrated in the very common moss *Ceratodon purpureus* in which apparently Correns[90] claimed to have seen them on one occasion.

It remains to touch on a few general matters relating to vegetative propagation in bryophytes. One is the question of distribution of

---

* See Crundwell & Nyholm[95] for some hitherto overlooked species in the *B. erythrocarpum* complex.

gemmiferous species in the different major groups. A survey of the British bryophyte flora shows that species which produce true gemmae are quite unevenly distributed taxonomically. Thus, gemmae are found in under 10% of Marchantiales, in just under 20% of Metzgeriales, whilst nearly 40% of British Jungermanniales (leafy liverworts) are known to produce them. The leafy liverworts are thus the group with the strongest tendency to form gemmae, if one uses the term in a wide sense. Schuster,[384] however, alludes to one- or two-celled structures as alone constituting 'true gemmae', so that he is able to draw a sharp distinction between the many families of Jungermanniales where these minute propagules are found and others, such as Lejeuneaceae, Frullaniaceae and their allies, where these 'true gemmae' are entirely replaced by deciduous organs or by structures akin to the one we have described in *Metzgeria*. He stresses that a particular kind of propagule will tend to occur throughout any one genus, or sometimes family, of liverworts.

Turning to the question of the distribution of gemmiferous species by habitat, there seems little doubt that the conquest of two particular habitats by bryophytes has been associated with abundant and efficient gemma production. These are (1) the trunk, branches and leaves of trees, (2) soil surfaces, for example those of greenhouse pots, fallow fields and kindred habitats connected with the cultivation of the land by man. In moist, oceanic climates such as that of south-west Ireland, an abundance of bryophytes can be seen on the trunks and branches of trees. Species of leafy liverwort bulk large there, whilst among mosses genera such as *Zygodon, Orthotrichum* and *Ulota,* all with some gemmiferous species, are likely to be among the most prominent in positions high above the ground. Gemma production must be a factor in the success of some of these species, whilst others like *Ulota crispa* are widespread because they fruit freely. In parts of the humid tropics minute liverworts abound on the surfaces of large, long-persistent leaves of forest trees. These epiphyllous forms constitute a highly specialised group and nearly all are Jungermanniales, especially of the family Lejeuneaceae. Many of the genera involved, whilst not forming gemmae in the narrow sense of Schuster, have the power to shed branchlets, leaves or portions of leaves and thus utilise these varied structures as propagules. One must not overstress this question of the supposed 'ecological advantage' conferred by vegetative propagation, for assessment of factors in competition is notoriously difficult. Dr E. W. Jones tells me (*in litt.*) that, in his view, copious and early production of spores is more important for epiphyllous Lejeuneaceae. Schuster is in no doubt, however, that on the more general level there is 'great selection pressure in favour of any device which will bypass syngamy'.

The implications of this statement would take us well beyond the scope of the present chapter.

As regards the agents of dispersal of gemmae, it is usually assumed that water is the most important: Brodie[49] has described a 'splash-cup' method of dispersal. Heavy rain must commonly be effective; but we may suppose that at various times the feet of animals and the wheels of vehicles will carry gemmae to new areas. Most types of gemma are unlikely to withstand drying, and wind will not therefore be important, except in areas of high humidity.

Enough has been said to show that vegetative propagation in bryophytes is a large and important topic. It is also one that has not lacked investigators, from the time of the classical morphological accounts to the modern period when the experimentalist is eager to make use of the special powers of bryophytes in this respect in his attempt to uncover some of the mechanisms of growth and differentiation. At the same time, we must not forget that in the lives of many bryophytes it is at most a secondary phenomenon. For the majority, in accordance with the accounts of the life-cycle in standard texts, the events which follow one another in unvarying alternation, generation after generation, are the union of gametes and the production of spores. With the first of these we are concerned in Chapter 8.

8

SEXUAL REPRODUCTION

An important event in the life of the gametophyte is the appearance of sex organs. The male and female organs, termed respectively antheridia and archegonia, are remarkably uniform in fundamental structure throughout bryophytes. They are ordinarily provided with some protection. The protective devices concerned, and the frequent aggregation of sex organs in specialised parts of the plant body, commonly render the fertile regions conspicuous, although the essential organs are themselves microscopic.

The antheridium is a delicate sac enclosing antherozoid mother cells and, ultimately, mature male gametes. In form it varies from subglobose, through ovoid to ellipsoid; the wall of the sac is almost always one layer of cells and the whole structure is borne on a stalk of varying length. The antheridium invariably originates from a single cell, normally superficial, but hypodermal in *Anthoceros*. A primary stalk cell is cut off early, whilst rather later the body of the antheridium differentiates as a result of periclinal cell divisions which separate the wall from the interior mass of antherozoid mother cells. Meanwhile the stalk region becomes multicellular.

In the Marchantiales, where these organs are sunk in characteristic chambers, the antheridium is ovoid, the stalk short and broad (Fig. 18E). In Metzgeriales the form (cf. *Pellia* (Fig. 18A)) is more nearly globose, but the stalk is still short. Typical leafy liverworts (Jungermanniales), however, have antheridia borne in the axils of concave leaves and the stalk is quite long (Fig. 18D). Also, as Parihar[320] points out for *Porella,* the wall may be locally two or three layers of cells thick. Mature antheridia of *Sphagnum* (Fig. 18C) resemble those of leafy liverworts in form, but those of Bryidae and Polytrichidae are much longer and narrower than anything seen among typical hepatics (cf. Fig. 18G,H). Each antheridium here grows for a considerable time by the activity of a clearly defined

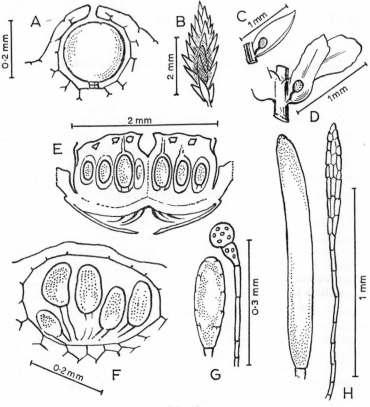

Fig. 18

The form of antheridia. A. *Pellia epiphylla* (vertical section), globose, shortly stalked, solitary, over-arched by wall of 'crater'. B–C. *Sphagnum rubellum*. B. Catkin-like (coloured) male branch. C. Antheridium, widely ovoid, long-stalked, in axil of bract (from B). D. *Diplophyllum albicans*, part of male branch, with single concave, deeply bifid male bract and antheridium. E. *Reboulia hemisphaerica*, median vertical section of male 'cushion', showing ovoid, shortly stalked antheridia in radial rows, the antheridial chambers open to exterior; on underside are ventral scales. F. *Anthoceros* sp., vertical section of antheridial chamber showing parts of five antheridia. G. *Funaria hygrometrica*, single antheridium (ellipsoid, shortly stalked) and paraphysis from male receptacle. H. *Polytrichum* sp., antheridium and larger (club-shaped) type of paraphysis.

apical cell. Although the body of the antheridium is thus elongated the stalk remains short. These moss antheridia vary much in size. For example, typical antheridia of *Funaria hygrometrica* attain a length of 0·25 mm, those of *Polytrichum commune* 1·5 mm. Those of *Mnium hornum* are intermediate between these two; whilst in plenty of mosses the antheridia are smaller than in *Funaria*. In the Calobryales, as Schuster emphasises, both the development and the mature form of antheridia are much more like those of mosses.

The archegonium is often, and fairly aptly, described as flask-shaped. The essential parts are two, a wide rounded venter which houses the egg (female gamete) and ventral canal cell, and a long narrow neck. The whole structure is commonly borne on a short stalk. The neck is a tube enclosing a core of intensely protoplasmic neck canal cells. These, together with the ventral canal cell, break down to afford access to the egg. There is thus a single, non-motile female gamete, adequately protected in the hollow of the archegonial venter. It may be mentioned that the distinction between the axial row of protoplasmic cells and the sterile wall cells is more fundamental than is that between the broad, basin-shaped venter and long, narrow neck.

Like the antheridium, the archegonium normally originates from a single superficial cell, but again *Anthoceros* is exceptional in that the resulting structure is not a free, elevated organ as it is in other bryophytes. Instead, as we noted in Chapter 2, the wall of the archegonium is here confluent with surrounding thallus tissue and the archegonium as a whole is remarkably small and short in the neck (Fig. 19D). The neck canal cells in *Anthoceros* are only four to six, whilst in most other liverwort groups they are more numerous, although only four occur in some Marchantiales (e.g. *Riccia*) and in *Sphaerocarpos*. Ten or more neck canal cells are found in typical moss archegonia, sixteen to forty in the exceptional liverwort *Haplomitrium* (Calobryales[385]). The separation of the wall from the central part of the archegonium is an important feature of early development, just as it is in the antheridium. The distinctive features of the female organ become apparent only as the venter enlarges to accommodate the big egg cell and the neck lengthens sharply. Also, the central protoplasmic cells are always few by comparison with the great number of antherozoid mother cells in the antheridium. The stalk is almost wholly lacking in many genera of Marchantiales and Metzgeriales. It is particularly long and massive in some of the Bryidae (Fig. 19F). An apical cell governs archegonial development in mosses as a whole, including *Andreaea* and *Sphagnum,* a fact which makes for relatively elongated archegonia throughout the class. Among liverworts, *Anthoceros* and *Haplomitrium* provide

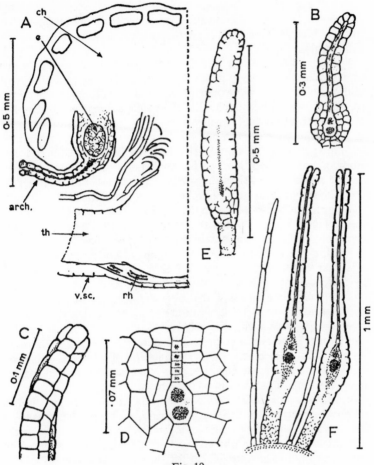

Fig. 19

Archegonia. A. *Preissia quadrata*. Part of vertical section of young carpocephalum, with archegonial venter containing very young embryo (stalk
of carpocephalum not visible). arch. archegonial neck; ch. chambered
dome of carpocephalum; e. embryo; rh. part of rhizoid; th. parent thallus;
v.sc. ventral scale. B–C *Fossombronia angulosa*. B. Archegonium removed
from apical part of thallus midrib, in optical section. C. Twisted neck of
same, surface view. D. *Anthoceros sp.* V.S. superficial layers of thallus with
immersed archegonium. E. *Hypum cupressiforme*. Abnormal sex organ,
intermediate between antheridium and archegonium (found in female
'inflorescence' of plant from Gau Graig, Cader Idris, July 1949). F. *Bryum*
sp. Two old archegonia, two paraphyses; note long stalks of archegonia.

the chief examples of divergent archegonial structure (in contrasted directions). In general there are six rows of neck cells in Marchantiales, five in Metzgeriales and Jungermanniales, and usually four in Calobryales.

We must now consider the morphological significance of the sex organs. We have already noted that both archegonium and antheridium achieve their maximum in size and complexity in the bryophytes. It is usually thought that their long subsequent history has been a path of gradual degeneration, with the antheridium ceasing to be recognisable as such among the gymnosperms and the archegonium persisting at that level, albeit in highly modified form. Both organs, indeed, tend to be smaller and simpler in structure among pteridophytes than they are in bryophytes. In the latter it is only the small, short-necked archegonia of some thalloid liverworts that approach pteridophyte archegonial structure, one of the closest resemblances being between the embedded archegonium of *Anthoceros* and that of such a pteridophyte as *Selaginella*.

Thus, if these characteristic sex organs arose at bryophyte level it is clear that they quickly achieved their peak of structural elaboration. If, as some believe, the bryophytes represent a downgrade step from pteridophyte ancestry, then this step, retrograde in some respects, brought with it an elaboration and diversification of sex organ structure. In any event, the relatively sudden appearance, in the evolutionary sense, of a complex and stereotyped organ such as the archegonium immediately raises the question of its origin. Whatever our views on bryophyte ancestry, it seems reasonable to look among algae for the ultimate origin of this organ.

In such a genus as *Ectocarpus,* among the brown algae, we find an ellipsoid structure (in origin a single cell), which produces gametes that are functionally either male or female. To postulate this as the ultimate origin of the antheridia and archegonia of bryophytes is to imply a common ancestral form for these two kinds of sex organ. This seems reasonable because, although among bryophytes so distinct at maturity, antheridia and archegonia show many resemblances in their young stages. This is especially true of the Calobryales. Concerning the extent of modification entailed, Parihar[320] refers one back to an early paper by Davis.[96] To transform the 'plurilocular sporangium' (gametangium) of *Ectocarpus* into an antheridium, one requires little more than the addition of a delicate 'jacket' of sterile cells. Transformation into an archegonium is a bigger step, for it must involve (besides the formation of a 'jacket' layer) a profound change of form in the organ as a whole and the pre-selection of a single internal cell as the functional female gamete. The existence of abnormal structures, intermediate between antheridia and archegonia, could furnish a useful line of evidence on this whole question. There has indeed been no lack of examples.

Observations on intermediate structures go back at least to those of Lindberg on a species of *Hypnum* in 1878. At intervals in the years that followed others were brought to light, by Hy[201] in *Atrichum undulatum,* by de Bergevin[98] in *Plagiothecium silvaticum,* by Holferty[188] in *Mnium cuspidatum.* Bryan[51] reviewed the earlier literature and added his own striking observations on *Mnium medium.* In this synoecious species he found that each of two populations, situated sixty miles apart, had numerous abnormal organs intermediate in various ways between antheridia and archegonia. Bryan figured twenty of these. Some had what were clearly antherozoid mother cells within the stalk of an otherwise fairly normal archegonium. Others were substantially antheridia, but had an apparent egg and ventral canal cell amongst the mass of normal antherozoid mother cells. We now know that the usual haploid chromosome number in the genus *Mnium* is six, but that *M. medium,* in common with certain others, has the chromosome status n = 12. It seems that polyploidy has here brought with it the situation in which the two kinds of sex organs are borne on a common receptacle. It is perhaps not surprising therefore that some intermediates should occur. They may be more frequent than is realised. I once came across what were obviously various intermediate types of sex organ (cf. Fig. 19E) when examining herbarium material of *Hypnum cupressiforme.* Bryan considered that such occurrences give ground for the belief that the two kinds of sex organ are homologous structures.

Haupt[171] in a study of the liverwort *Preissia quadrata* found that, although dioecism is the rule, about 1% of all plants seen were monoecious. Some monoecious plants carried bisexual receptacles in which there were a few abnormal, intermediate sex organs. Haupt figured an organ which looked as if it would have completed development as an egg-producing structure but which lay in a typical antheridial chamber. Again, on mainly female receptacles he found some undoubted antheridium-like structures. Haupt concluded that male and female receptacles of Marchantiales had had a common origin phylogenetically; also that antheridia and archegonia were homologous organs. Records of intermediate or otherwise abnormal sex organs continue to appear in the literature (cf. Denizot,[101] Bhandari and Lal[28]) and the conclusion as to common origin, at some time in the very remote past, seems inescapable.

We will now turn to the arrangement and grouping of sex organs. In many genera of both mosses and liverworts dioecism is found and evidence of this is usually provided by plants of the two sexes growing in completely separate patches. Sometimes separate male and female plants appear to exist but they have in fact sprung from a common protonema. This condition may be termed 'pseudo-dioecism'. In many instances of true dioecism plants of the two sexes differ greatly in size and appearance. In *Buxbaumia* among mosses and *Sphaerocarpos* among liverworts, as we have seen, the male plants are minute. Quite often (cf. Gemmell[140]) dioecism seems to have given place to monoecism in the evolutionary history of a genus. Thus *Mnium medium* is a monoecious species in a genus where the majority are dioecious. Noguchi and Osada,[308] in a review of the Japanese species of *Atrichum,* have revealed a similar situation, the majority of the species being dioecious with a haploid chromosome number of 7; but two varieties constitute exceptions, *A. undulatum* var. *undulatum* being monoecious ($n = 14$) and *A. undulatum* var. *haussknechtii* actually having the sex organs together in a common inflorescence, the haploid chromosome number here being 21. A comparable situation exists elsewhere, for example in *Pohlia* and *Hypnum,* as indicated by Lewis[260] in an important paper which deals with some of the genetical implications of these facts. Lewis, citing Yano,[473] also instances several species in which dioecious and monoecious strains are known. In each case the haploid chromosome number of the monoecious strain is double that of its dioecious counterpart. This suggests that monoecism is ordinarily a derived condition.

Where the two kinds of sex organs occur on the same plant they may develop in separate receptacles (loosely called inflorescences); the plant is then autoecious. In some monoecious species the apical cell cuts off antheridial initials closely followed by archegonial initials so that the two kinds of sex organ are separated only by one or two leaf insertions; this is the paroecious condition. Finally, in a synoecious species the two are intermingled in a common receptacle. Differences in 'inflorescence' form an important criterion for the separation of closely related species in many groups of bryophytes. Notable examples are found in *Bryum* and *Pohlia*. Yet some species, even in these genera, are conceded to be variable in this respect, and such differences may prove to be less fundamental than they were at one time thought to be.

As regards the arrangement of antheridia, thalloid and leafy plant bodies pose two quite different problems. In thalloid liverworts the sinking of antheridia in 'pits' or chambers not only provides efficient protection but also allows for a mechanism of quite violent discharge of antherozoids. The male discs or 'heads' (antheridiophores) of many Marchantiales show this well. At the time of discharge the wall cells of the antheridium become distended and their abutment against the walls of the chamber brings pressure to bear on the contained mass of antherozoids. The key to the process lies in the formation of abundant mucilage: also in the differentiation of a cap cell (or group of cells) which bursts to release the male gametes. Cavers[70] pointed out that in *Conocephalum conicum* this discharge was explosive, the antherozoids escaping in fine jets capable of reaching a height of over 5 cm. A measure of grouping of antheridia also prevails in many Metzgeriales (cf. Chapter 2), but in *Fossombronia*, by contrast, they occur scattered and unprotected, along the upper surface of the stout axis (as do the archegonia). According to Schuster, this is the more primitive condition, whence the embedded state has been derived.

The Anthocerotae have already been noted as unique in the hypodermal origin of their antheridia. As a result of a process of proliferation the resultant chambers, at least in *Anthoceros,* commonly contain two to five antheridia, often many more (cf. Proskauer[343]). In the often tiered wall cells of the antheridium in *Anthoceros,* moreover, green plastids are conspicuous. Later they become orange, and colourful patches marking male receptacles are then obvious to the naked eye. A similar sequence of pigmentation is not uncommon in the wall (jacket) cells of moss antheridia (cf. *Funaria*). Schuster[384] reminds us that an orange-yellow pigmentation also distinguishes the mature antheridia of *Blasia, Fossombronia* and *Haplomitrium,* among Hepaticae, but that in most liverworts a pale greenish hue prevails.

In the Jungermanniales (leafy liverworts) the slender-stalked, subglobose antheridia arise singly, or in groups of two or three, in the axils of specialised concave leaves. The whole male shoot is often catkin-like in form, and quite different from the purely vegetative shoots. The exceptional size of these catkin-like shoots furnishes a useful specific character in the bog liverwort, *Cephalozia macrostachya.* A similar arrangement exists in *Sphagnum,* where the catkinate male branches are often conspicuous too for their highly coloured leaves (Fig. 18B). In all these examples the protection of the antheridium is only such as can be conveyed by the concavity and close proximity of the leaves (male bracts) concerned.

In *Funaria hygrometrica* (Bryidae), the antheridia are densely packed in a male receptacle where the sterile hairs, or paraphyses, have also an important part to play. Antheridia may be found in many stages of development and a single receptacle of *Funaria hygrometrica* is thus probably capable of releasing a succession of viable male gametes over a period of at least several weeks. It is easy to appreciate how the paraphyses in this species, with their swollen sub-spherical terminal cells (Fig. 18G), can be of value in protection, moisture conservation, building up pressure to make possible efficient discharge of antherozoids, and (to a limited extent) in photosynthesis. The terminal cells of the paraphyses meet over the antheridia. Both terminal and subterminal cells are rich in chloroplasts. Lorch[267] considered that their principal significance lay in conserving moisture around the antheridia.

Paraphyses usually form simple filaments which reach their broadest point above the middle, though not always (as in *Funaria*) in the terminal cell itself. Also, I have examined paraphyses in *Mnium, Bryum, Pohlia, Philonotis* and other genera without finding any that are as richly chlorophyllose as those of *Funaria hygrometrica*. In *Polytrichum* some of the paraphyses are more elaborate, being several cells wide in their broadest part (Fig. 18H). Always the longest paraphyses seem to overtop the antheridia so that the essential organs are in a sense embedded. Sometimes, e.g. *Philonotis,* the special form of associated leaves (perigonial bracts) makes for additional protection and the whole mass of antheridia and paraphyses may appear as if sunk in a kind of hollow. The antheridia in such a cluster may number several hundred.

It has sometimes been suggested that more use might be made of paraphysis structure as a taxonomic character, but it is seldom that really clear-cut differences exist between the paraphyses of two closely related species. Dixon[107] emphasised how valuable this character can be in separating non-fruiting plants of *Splachnum sphaericum* and *Tetraplodon wormskjoldii*, but even here there are other readily seen differences. In most mosses antheridial production ultimately uses up the apical cell, but *Polytrichum* is an exception. Here proliferation through the old antheridial receptacle is well known, and one commonly sees examples in which several successive years' male receptacles have arisen at short intervals in this manner.

The position and arrangement of archegonia have already received some attention because of their connection with the subsequent development of the sporogonium. For example, on p. 27 a brief survey was given of the situation obtaining in diverse thalloid forms. Thus, solitary archegonia, lacking a clear-cut arrangement, are found in *Riccia,* in *Sphaerocarpos* and again in *Anthoceros*. By contrast, in the mature 'mushroom-head' (archegoniophore) of the highly specialised *Marchantia polymorpha* we find a regular sequence of archegonia in a series of radial lines, the youngest being nearest to the stalk of the archegoniophore. Immersion in thalloid tissue tends to be most pronounced where ancillary protective structures (involucre or pseudoperianth) are lacking. In *Fossombronia* (Metzgeriales), however, the archegonia lack all protection except that provided by the crowded younger leaves; yet they are in no way embedded and are easily observed.

Throughout most Jungermanniales a small group of archegonia forms and it is characteristic of this order of liverworts that the apical cell is used up in the formation of the final one (acrogyny). In the very large family Lejeuneaceae only one archegonium occurs—clearly a derived state attained by these highly specialised leafy liverworts. The situation in *Sphagnum,* with its terminal group of three, is very similar to that in most Jungermanniales. Again the apical cell is used up in the production of the last-formed, centrally placed archegonium. In Bryidae the situation is not very different but it is usual for larger numbers to be involved. In species of *Bryum,* and various other genera besides, one commonly meets with a cluster of twenty to thirty archegonia, close-packed in the 'inflorescence' and accompanied by slender paraphyses. Receptacles in many pleurocarpous mosses, however, tend to be far less bulky; but it is not easy to come upon a detailed statement regarding the precise number of archegonia forming in the receptacles of different orders and families of mosses.

Two special features distinguish archegonia in which the ovum (or egg cell) is ready for fertilisation. One is the breakdown into a swelling slimy mass of the neck canal cells and ventral canal cell, with the consequent separation of the cover cells. The other is the secretion of special chemical compounds which serve to bring the motile male gametes not only to the neck of the archegonium but down the narrow passage that leads to the egg. According to Parihar,[320] the secretion in *Riccia* may consist of proteins and inorganic salts of potassium. The same author implies that most other bryophytes display a similar mechanism, but there seems to be little precise information. Various authors have stated that cane sugar is the active substance in the moss *Funaria*. Showalter[397] pointed out that in *Riccardia* the disorganised canal cells were not responsible for the secretion, although some modern authors (cf. Parihar[320]) continue to imply that they are, at any rate in most bryophytes. Meanwhile the ovum (or egg cell) has undergone a specialised maturation process. Diers[105] showed that in *Sphaerocarpos donnellii* this entailed an eightfold increase in cell volume and a fifteen-fold increase in nuclear volume. The numbers of mitochondria and plastids were also greatly increased. A very remarkable feature of this maturation process is the production of extensive evaginations by the nuclear membrane (Diers[105]).

Where the two kinds of sex organ lie near one another the onset of wet conditions will ordinarily supply sufficient water to enable male gametes to reach the necks of archegonia. In some monoecious and all dioecious species, where male and female receptacles may be widely separated, rain splash and the visits of microscopic animals such as mites have been recorded as mechanisms of transfer. Considerable light, however, was thrown on the whole subject by the researches of Muggoch and Walton.[298] These authors referred to the earlier observation of Showalter[396] that if a large number of antherozoids (of *Riccardia pinguis*) were placed at one end of a small pool of water 1 cm long by 0·5 cm wide, nearly all remained crowded at that end of the pool an hour later. Hence one must not lean too heavily on the swimming powers of the motile male gamete. Moreover, Muggoch and Walton remind us that it is usual for antherozoids to leave the antheridium in a compact mass, still enclosed within their mother cells. Even in *Conocephalum* the male gametes are still within delicate membranes when they are ejected in violent spurts.

The important contribution of the above authors was to reveal the part played in the process by a spreading agent that was capable of drastically lowering surface tension. It operated after the release of the mass of 'sperm cells' and before the onset of the free swimming phase. Fat was probably responsible, for this was detected in all antheridia which showed this property but was lacking in *Conocephalum* and *Sphagnum*, two genera in which results were negative. In a later note Walton[441] pointed out that in *Pellia epiphylla* only fifteen seconds were required for the surface spreading, but a further fifteen minutes might elapse before the free-swimming antherozoids were fully released from their mother cells. By means of this important mechanism they had already been conveyed almost to the necks of the archegonia.

We may now examine the final stages of development of the cells within the antheridium. The terminology adopted by Smith[403] and Parihar[320] is not wholly satisfactory, since it results in the term 'androcyte' being used for something different from the 'spermatocyte' of Muggoch and Walton[298]. It is better for our purpose to avoid unnecessary terminology. In all nearly mature antheridia there are numerous small cubical cells. Original figures by Ikeno[202] for *Marchantia polymorpha* and by Black[35] for *Riccia frostii* (widely copied since) imply that, in these examples at least, the contents of the final generation of cubical cells (spermatocytes) divide diagonally into two, without the formation of a cell wall. Each resulting unit, or spermatid (androcyte of Smith) then becomes transformed into a male gamete. One certainly sees these paired spermatids in longitudinal sections of 'late stage' antheridia of *Pellia,* but there is some doubt as to how widely such a division into pairs occurs in the various groups of bryophytes. Few mosses seem to show this feature (cf. Wilson[468] and other later authors).

I am indebted to Dr J. G. Duckett for help in the preparation of an up-to-date account of the transformation of spermatid into spermatozoid (last generation of androcytes into antherozoids). The following changes take place (1) The cells become rounded, then comma-shaped and finally elongated and coiled; (2) early in this metamorphosis a new structure (the blepharoplast) appears in the cytoplasm; (3) this blepharoplast becomes long, narrow and thickened at its anterior end, whence two backwardly directed flagella grow out; (4) during the metamorphosis the nucleus itself becomes spirally coiled, with its chromatin so condensed as to form almost a 'solid rod of DNA'.

The above is only an outline. Indeed, an entirely new light has been thrown on the structure of the spermatozoids by a long series of electron microscope studies. An unsuspected wealth of detail emerges. Manton and Clarke[283] made the first stride forward when they revealed the '9 + 2 strand' flagellar structure in the spermatozoid of *Sphagnum*, a 'pattern' which has since been shown to prevail, not only in other groups of bryophytes but also in *Equisetum* and in both ferns and cycads. Subsequent work has been concerned principally with the fine structure of the body of the spermatozoid itself. Sato[373] (also using the electron microscope) alluded to this as dumb-bell-shaped (in *Anthoceros*), and he recognised too, in all eight bryophytes examined by him, an apparently new structure which he called the 'filamentous appendage'. Manton[282] referred to the same structure as a 'fibrous band' and, more recently, Carothers and Kreitner[67] used the term 'spline' for it. These authors showed that it was composed of a band of up to thirty micro-tubules (each 250 Å in diameter) and formed part of the blepharoplast. Closely assocated with both nucleus and flagellar bases, it is not confined to bryophyte spermatozoids and is thought to convey structural stability to these naked motile cells.

Further complexities in blepharoplast structure continue to come to light. For example, the thickened anterior end consists of several distinct organelles in close juxtaposition. The outermost of these are two centrioles. They arise (apparently *de novo*) early in the metamorphosis, and it is from them that the flagella originate (Paolillo,[319] Carothers and Kreitner[67]). Beneath these is the band of microtubules; and there are other structures too. Most of them are also found in the spermatozoids of other archegoniates, but the 'limosphere' or 'apical body' appears to be unique to bryophytes. It consists of a plastid surrounded by a large mitochondrion. With all this hitherto unrecognised complexity of ultra-structure, the spermatozoid can justifiably be claimed as 'the most highly differentiated of all bryophyte cells' (Duckett, *in litt.*).

Everybody can examine the electron micrographs made available by the above investigators, but the ordinary student, for the most part, will have to be content with first-hand experience of a different order. In examining living material, for example of *Funaria*, he will find it is often possible to see for himself the gradual extrusion of the greyish mass of mother cells, and under higher magnification, the coiled form and rapid movements of the male gametes.

Union of the spermatozoid (antherozoid) nucleus with that of the ovum completes the process of fertilisation. Andersen long ago noted that in diameter the egg nucleus far exceeded that of the male gamete. Diers[105-6] has observed, in *Sphaerocarpos*, that a number of spermatozoids enter each archegonium. The details of the union of nuclei, however, await study by modern techniques. Thereafter one is concerned with events in the new sporophyte generation.

In all ordinary cases the successful fertilisation of one archegonium implies the loss of function of those that remain. Sometimes, after gametophyte tissue has grown to accommodate the increasing bulk of the sporophyte, these abortive archegonia may be seen in odd positions. In certain species of moss, however, multiple fertilisations are normal and two or more sporophytes arise within a single cluster of perichaetial leaves. *Dicranum majus* and *Mnium undulatum* are good examples. Longton[263] has given a useful survey of this condition (polysety) in British bryophytes. Entirely different again are the cases reported from time to time (Gyorffy,[162] Watson[443]) where it appears that a single fertilisation has occurred but, owing to subsequent abnormal behaviour of the apical region of the young sporophyte, a monstrosity results and two capsules stand united at the end of a single seta. Udar and Chandra[428] have described what is almost certainly a much rarer condition still, namely true polyembryony. Thus, in the liverwort *Mannia foreaui* they found that the existence of two egg cells within one archegonium resulted in the development of two superimposed embryos in a common venter. These authors allude to some earlier examples and to Land's general review of the subject.[245] Some examples of double sporophytes could of course be explicable on this basis.

In Chapter 6 reference was made to the work of Greene and Greene[152] on 'maturation cycles'. There has, indeed, been considerable interest in the past decade or so in the whole subject of the onset of the reproductive phase in bryophytes and the factors which govern the process, quite apart from the time sequence of events in particular examples. Since gametangial development and subsequent sporophyte growth are inseparably linked together and since the 'factors' investigated are those (light, temperature, etc.) associated with other physiological processes, this subject is most suitably considered in the next chapter where some aspects of morphogenesis, anatomy and physiology will be reviewed. In the present context, however, it is worth noting that the onset of sexual reproduction is usually far from rapid. Thus, Longton and Greene[265] point out that the time required for the maturation of antheridia varies between a few weeks and eleven months in different species of moss.

9

MORPHOGENESIS; ANATOMY; PHYSIOLOGY

The three approaches to the study of bryophytes which are to be considered in this chapter are all necessary for the fuller understanding of the living plant. Starting from different standpoints and employing diverse techniques, they converge in throwing light on the development and functioning of the plant as a whole. Morphogenesis emphasises the unfolding of the structural pattern and seeks to understand the causes which promote the onset of each new phase in development. Thus attention comes to be focused on certain stages, for example the onset of bud formation on the protonema of a moss or the first signs of differentiation in the very young sporophyte. A grasp of bryophyte anatomy is essential because it is only in terms of anatomical structure that the developmental changes can be observed and followed. Furthermore, the structural adaptations revealed may well provide clues to physiological processes. Finally, the more important physiological activities of bryophytes are today being submitted to the rigours of precise experimental study. Hence the decision to include some review of these three big topics in a single chapter. Thereby we may be enabled to understand more clearly what follows in Chapter 10 under the head of bryophyte ecology.

First we must examine the very young stages of bryophytes, which are seen immediately after the germination of spores. Fulford, besides giving accounts of certain chosen examples, has ably reviewed the position as a whole in the leafy liverworts.[132] She points out that as long ago as 1862 Hofmeister recognised three distinct patterns of early development among liverwort sporelings. Leitgeb[256] extended

these to four, whilst Goebel[148] was the first to describe some of th
distinctive patterns found in the family Lejeuneaceae. Fulford,[132]
who has added much to our knowledge of this field, recognises
ten structural patterns of the first phase after spore germination.
These, with genera exemplifying them, are: the filamentous type
of *Cephalozia,* the cell body of *Nardia,* the disc of cells found in
Radula and the ball of cells found in *Frullania*; these being followed
by a further six patterns prevailing in different members of the
Lejeuneaceae. In a more recent study Nehira[302] has recognised
twenty-four different sporeling patterns (seven of them in Metzgeriales) and attempted to draw some phylogenetic conclusions.
An almost globose sporeling in *Calobryum* (*Haplomitrium*) reminded
him of *Megaceros* and was deemed to be primitive. Both Schuster[384]
and Nehira[302] have warned against too rigid an acceptance of
Fulford's ten 'types', emphasising that environment can exert a
profound effect in many cases, as Chalaud[76] long ago had shown.
A sharp end to the 'protonemal' phase is not always detectable in
liverworts and it is well to remind ourselves that in this group (unlike
mosses) one 'protonema' gives rise to one plant. Two further stages
ordinarily intervene before the appearance of stem and leaves of
adult form. These are (1) the shoot with primary leaves (which may
be unicellular, may consist of a cell row, or may be broader at the
base) and (2) the shoot with juvenile leaves, which may in varying
degrees approach the form of adult leaves.

Where the adult form is a thallus the early stages are on the whole
more contracted, and events proceed more directly towards the final
end to be achieved. Inoue,[206] and later Nehira,[301] have looked at
examples from various families of Metzgeriales. In most instances,
by the time a few rhizoids have emerged the young plant presents an
irregular-shaped cell mass. Only in some species of *Riccardia* is a
fairly well defined filamentous stage observed. *Pellia* appears to be
unique in the group in the advanced, multicellular state that is
achieved within the spore wall. In it polarity is early acquired and
the basal cell may be distinguished by its less dense cytoplasm and
few chloroplasts. From this cell the first rhizoid grows out and in
due course one of the wedge-shaped cells cut out at the opposite
pole will become the functional apical cell of the young plant. In
all the other types recognised by Nehira in this order the start of
germination is quite different, the spore often first of all enlarging
to fully twice its original diameter.

The early stages of the gametophyte in the Marchantiales have
been investigated by O'Hanlon,[313] Menge[288] and others. More
recently, Mehra and Kachroo[287] have examined members of the
Rebouliaceae and Kachroo[226] has described the early development

vophylla. In the latter he recognises four stages: ... e of a germ papilla; (2) the formation of a row of ...be; (3) expansion of this into a germ plate; and ... in which the apical cell is ultimately cut out. ...amentous young stage, albeit one that is quickly ...ued; and Kachroo remarks that secondary filamentous germ tubes may form in conditions of feeble illumination, just as they do in the Rebouliaceae. Inoue,[206] and later Schuster,[384] have seen the very distinct sporeling pattern of *Monoclea* as an argument for its exclusion from the Marchantiales. It would seem that in these thalloid forms, as in the leafy liverworts, there is an underlying pattern of development but some plasticity in the face of unusual conditions.

Among mosses thalloid juvenile stages are rare, and most commonly a filamentous phase (the protonema) gives place ultimately, and abruptly, to the leafy shoot system of the gametophore. Reference has been made earlier to the so-called thalloid protonema of *Sphagnum, Andreaea* and a few others, and we need not dwell on these exceptional cases here. Instead we will consider some recent work on the freely branched filamentous protonema of ordinary mosses. *Funaria hygrometrica* has been a frequent subject of study. The principal aims have been to distinguish successive developmental phases, and to suggest if possible what mechanisms control the onset of each new phase. This is the special province of the experimental morphologist, who must also often be something of a physiologist as well. Unfortunately, there is as yet no full agreement as to the stages in protonemal development to be recognised, and this despite much critical work by investigators in several countries.

The central controversy concerns the question as to whether, in such a moss as *Funaria hygrometrica,* one can recognise two clear-cut stages in protonemal life and form, the stages that have been named chloronema and caulonema. This idea of a clear two-phase development was put forward by Sironval[400] and has since been strongly supported by Bopp.[40] The latter author lists six morphological characters in which the first-formed chloronema differs from the caulonema which succeeds it. Among other things, the chloronema shows irregular branching, no oblique cross walls, colourless cell-walls in general and numerous, evenly distributed chloroplasts which are mostly nearly circular in outline. At a certain stage (after about twenty days according to Sironval) this is succeeded by caulonema, which Bopp distinguishes by its very regular lateral branches, oblique cross walls, a tendency for cell walls to be brownish and chloroplasts fewer, spindle-shaped and less evenly distributed. He adds two further differences concerning the nuclei. In a subsequent important review of many aspects of morphogenesis in mosses Bopp[41] has upheld his verdict that these two phases are truly distinct. Moreover, it is only on branches of caulonema that buds can be formed.

Clearly, if this is true, a study of the conditions controlling the onset of caulonema will be important; and in this connection Bopp has pointed out that low temperature, submersion and low light intensity may each independently delay or prevent the establishment of caulonema, and hence of buds. The quality of light is also said to be important. Bopp goes on to explain that these two phases differ physiologically. Caulonema, unlike its forerunner, will show a negative response to unilateral illumination; and two unconnected caulonema branches growing adjacent to one another were found to impede each other's growth, as the result of an unknown inhibitor being secreted. Bopp claims to have recognised these two phases in other genera, for example in *Fissidens, Barbula, Splachnum* and *Bryum*. He believes that the two decisive facts which prove caulonema to be a distinct stage of differentiation are (1) that it requires definite conditions for its appearance, and (2) that its physiological reactions are different from those of chloronema.

Both van Andel[430] and Allsopp and Mitra[5] have been unable to confirm the above interpretation of protonema as consisting of two clear-cut phases. Even Kofler,[239] in her exceedingly thorough investigation of the early gametophyte stages of *Funaria hygrometrica in vitro*, remained very imperfectly convinced. Nehira,[303] more recently, was able to demonstrate distinct chloronema and caulonema in *Pleuroziopsis ruthenica*, but found no evidence for it in *Fissidens heterolimbatus*. The third Japanese moss which he studied was *Schistostega pennata*, which has long been known to have a specialised type of protonema incorporating a certain proportion of lens-shaped cells with peculiar light-reflecting power. Some of the discrepancies in interpretation of protonemal morphology no doubt arise from the fact that protonema is both subject to considerable genetic variation and also highly plastic in its reaction to different environmental conditions. Kofler indeed claims that protonema brings out genetic differences more strongly than do the gametophore and sporophyte, where the environment has had more time in which to exert an effect. Concerning plasticity, Bopp found that if isolated cells (or short fragments) of caulonema were taken apart, they grew but reverted to chloronema. Kofler found that the presence of ammonium nitrate resulted in a slower protonemal growth throughout, and an abrupt falling off altogether after about three weeks. She found other factors, including impurities in the agar, which could exert an influence.

The crucial point in protonemal development is that of bud initiation, with the formation of a tetrahedral apical cell whence a leafy shoot can arise. Not surprisingly, therefore, many workers in recent years have studied the effect of varying experimental conditions on this point. Among them, during the 1950s, were Gorton and Eakin[149] in America and Mitra and Allsopp[291] in Britain. Gorton and Eakin[149] found evidence of an unknown inhibitor, which prevented further elongation of protonemal filaments after a certain stage had been reached. They saw this stage as closely connected with the onset of bud formation, for which, however, they considered that two further factors were important (1) the building up of an adequate nutrient store and (2) the supply of certain growth substances. They worked on *Tortella caespitosa*. Mitra and Allsopp[291] (partly in collaboration with Wareing[292]), studying *Pohlia nutans,* concluded that, although a certain minimum concentration of sugar in the protonema may be essential for bud formation, it is probable that some more specific substance, synthesised by protonema only in the light, is required.

In the last decade there have been numerous experimental morphologists investigating the effects of particular factors on bud formation, as also indeed on spore germination and protonemal structure. Light of differing wavelengths, X-rays, glucose and other sugars, and various growth promoting substances have all been studied. Thus, Oehlkers[312] found that X-ray treatment of swollen spores of *Funaria hygrometrica* induced variants in which particular parts of the 'transition from protonema to moss plants' were always characteristically altered. Glucose, applied to X-ray mutants, or to dark-grown protonema, can facilitate bud formation. Kinetin has been widely shown to increase the number of buds formed (on a protonema in the light) and to induce their formation even in dark-grown protonema (cf. Gourgaud,[150] Chopra and Gupta[79]). Mitra and co-workers showed that blue light can inhibit bud formation but that kinetin can quickly reverse this effect. According to Bopp[42] (whose survey of the subject made in 1968 is a rich source of information), gibberellins, in general, act to increase the number of buds whilst indoleacetic acid in low concentrations can occasionally have a similar effect. The whole question, however, is greatly complicated by the varied effects which these and other substances can exert in different combinations; also by such endogenous regulators as the 'factor H' reported by Klein[235] in 1967 which in high concentration apparently inhibits caulonema but in low concentration promotes bud formation. Without going into further detail we may quote, by way of summary, the apt words of Bopp.[42] 'The object of nearly all these experiments is to comprehend the switching mechanisms which, during the control of differentiation, become activated by external factors or inducing substances'. As yet we comprehend them only in part.

The studies of Meusel[289] have shown that even when the first leafy shoots have grown from the initial crop of buds the gametophyte is still far from mature. Several successions of immature leafy shoots ordinarily follow before shoots bearing sex organs are produced. Thus, the position is perhaps not very different from that already outlined for leafy liverworts; but there the successive phases are more minute and more rapidly passed through. Reflecting on Meusel's observations of the first-formed leafy shoots in mosses, one feels that they have not received all the attention that they merit. There is even the possibility that some of the taxonomist's 'varieties' may be referable to immature states of a species.

Eventually shoots will arise which are capable of bearing sex organs. The actual appearance of antheridia and archegonia, however, is under the control of various factors. As Benson-Evans[24] has shown, day length is undoubtedly of overriding importance, although temperature can exert at least a modifying influence in particular cases. Of fourteen bryophytes initially studied by Benson-Evans, four Marchantiales and six Jungermanniales were shown to be 'long day plants' so far as sex organ initiation was concerned. *Riccia* species, *Anthoceros laevis* and *Sphagnum plumulosum* were short day plants, whilst *Polytrichum aloides* appeared to be neutral. Later, investigating common mosses, Benson-Evans and Brough[25] found that the sex organs of *Atrichum undulatum* and *Mnium hornum* matured in a spring to summer period, those of *Eurhynchium praelongum* and *Funaria hygrometrica* in autumn or winter.

The sporophyte presents its own riddles to the experimental morphologist. The central question is what determines the unfolding of a completely different structural pattern as soon as fertilisation has been effected. As Bopp has remarked, we are still far from a full answer to this question, but much interesting and suggestive work is in progress. It is certainly not the fact of diploidy in itself, as ample experiments have demonstrated. It has been known since 1878 (Pringsheim[338]) that chopped up (diploid) seta tissue will regenerate protonema, thereby (as Bauer puts it) showing a 'collapse of the level of differentiation' that one might have thought to be characteristic of diploid tissue. Indeed, all three 'levels'—protonema, leafy shoot, and leafless axial sporophyte form—can occur in both haploid and diploid material. Bauer[21] has himself paid particular attention to the potentialities of chopped up sporophytes and has shown that, for the regenerant to go on to produce a new sporophyte apogamously, the cultured portion must come (a) from a young spindle-shaped sporophyte and (b) from that part of it which was destined to develop into the proximal part of the capsule. Completing the picture, as it were, Lal[244] has drawn attention to the spontaneous production of abundant apogamous sporophytes in haploid material of *Physcomitrium coorgense* cultured by him. They would arise on stems or on the sides of archegonia; sometimes successive phases of unorganised callus and apogamous sporophyte production would follow one another.

Apogamous sporogonia have been described in various species by a number of other morphologists (cf. Bauer,[21] Lazarenko,[252] Chopra & Rashid[80]), back to the time (1935) when Springer[406] first drew attention to this phenomenon. Recently, Hughes[197] has been interested in a comparison of normal sporophytes and apogamous (also diploid) sporophytes in their differing reactions to various environmental factors. In order to account for differences seen he found it necessary to postulate a controlling influence exerted by the leafy gametophyte on the developing sporophyte. The apogamous sporophyte, not being thus 'underpinned' by a leafy gametophyte, behaved quite differently. Thus, to some extent at least, Hughes sees the unfolding of the complex and characteristic structure of a moss capsule as under direct gametophyte control. Bopp[41] had earlier brought a different light to bear on this problem when he emphasised the importance of the intact calyptra as a control. Premature removal of the calyptra resulted, in different cases, in rapidly increased cell divisions or gross cell enlargement, or both. The outcome was the formation of a poorly differentiated monstrosity. This demonstration of the significance of the calyptra is in keeping with another important fact pointed out by Bopp; namely that, although the apophysis is the first part of the sporophyte to swell, the apical regions are the first to *begin* development and they act as a kind of 'organiser' for the rest.

We may turn next to the anatomy of bryophytes, in the sense of the kinds of cells and tissues displayed by the group. This study, which is receiving some attention at the present day, not only has an important bearing on our understanding of physiological processes, but also can on occasion provide valuable evidence on questions of evolution and taxonomy. Although in approach and objectives 'physiological anatomy' and 'systematic anatomy' are very distinct, it sometimes happens that the discovery of a particular piece of information can be of interest to workers in both spheres. The bryological anatomist does well to remember this. He also has to remind himself that features which he sees in the gametophytes of mosses and liverworts can be compared only in a rather indirect way with anatomical features that are familiar in vascular plants. For the latter are developments of the sporophyte generation. If we are to look among bryophytes for their strict counterpart (or homologue) we must seek it in capsule, seta and foot. It will be convenient to begin with this, the sporophyte phase of the life-cycle.

In the best-developed sporophytes (cf. *Anthoceros, Funaria,* etc.) we find epidermis, stomata, water storage tissue and a well developed chlorophyllose tissue within the capsule (cf. Fig. 25, p. 177). The chlorophyllose parenchyma, with its cells of specialised form and extensive air space system, shows in the apophysis of *Funaria* some striking resemblances to the spongy mesophyll of vascular plant leaves. Some elongated, probably conducting, parenchyma is commonly found in the columella region but the nearest approach to vascular strands, as understood among higher plants, consists of certain cells with spiral thickenings which Proskauer[346] has noted on the edge of the columella in *Dendroceros crispus*. In seta and foot the situation is quite different, the range of tissues smaller in accord with their less complex differentiation.

Lorch,[267] in his *Anatomie der Laubmoose,* alluded to the central strand that is sometimes present in the foot. This organ is flanked by parenchymatous cells of a highly specialised type (cf. p. 81) which must surely be involved in the transport of materials from gametophyte to sporophyte (cf. Chapter 6 and the physiological section of this chapter). The seta (Fig. 20D,E) is commonly stiffened by a peripheral zone of thick-walled cells, whilst in its interior lies an unmistakable conducting strand, with 'hydroids' and 'leptoids' (see later). The seta of the Australian *Dawsonia polytrichoides,* to judge from Lorch's figures, even has a sheath of cells strongly suggestive of endodermis. Although plenty of cells with very thick walls occur in the exothecium of aging moss capsules and a measure of rigidity prevails both here and in the seta, true lignification appears to be very largely lacking (cf., however, Siegel[398]). Cutinisation, on the other hand, is apparently widespread on the surfaces of these organs. Lorch quoted the earlier work of Strunk[414] in this connection, and some kind of cuticular covering must be responsible for the high gloss normally seen on seta and capsule. Without it the green capsule would be very vulnerable to desiccation.

Of greater ecological significance, because operative throughout the life of moss or liverwort, are the anatomical features of the gametophyte. Here again the incidence of cuticle is of interest. Buch[57] has shown that, on the whole, a delicate but perceptible cuticle is present on the leaves of those mosses in which the main water supply comes from below. In these, of which the genera *Bryum* and *Mnium* are good examples, the rhizoids are used to some extent to extract water from the substratum and it is then conducted up the stems to the leaves. They are called endohydric mosses. On the other hand, there are many mosses in which no cuticle can be demonstrated, and here the normal water supply is from above. In these instances the leaves often recover from the curled-up state remarkably quickly when they are supplied with water. Well-known examples are species of *Trichostomum, Tortella, Orthotrichum* and *Ulota*. They are called ectohydric. A further group, combining in some measure the characteristics of these two, is said to be mixohydric. The same division can be made among liverworts: most Marchantiales are endohydric, whilst leafy liverworts are mainly ectohydric.

It is clear that we have here two biologically contrasted groups that are likely to differ from one another in other anatomical characters, especially those that concern putative conducting tissues. At this point it will be convenient therefore to turn to the 'central strands' of some moss stems, and a few liverwort thalli, and the midribs that are found in the leaves of many mosses, where some interesting kinds of tissue differentiation are seen. 'Central strands' vary greatly in complexity. Among the simplest are those which have long been known in the thalli of such genera as *Pallavicinia, Symphyogyna* and *Hymenophytum* in the Metzgeriales. They have recently been re-examined in detail by Smith, J. L.,[404] who reminds us that Tansley and Chick[423] were the first to demonstrate (using eosin) the conducting function of these strands in *Pallavicinia lyellii*. Smith confirms this function and concludes that the fairly uniform tissue that characterises these strands is 'strikingly analagous to' the tracheidal xylem elements of vascular plants. The cells are long and narrow, with tapering ends, their walls bear numerous obliquely orientated pits and the loss of nucleus and protoplasm during development is clearly demonstrated.

The central strands in moss stems are made more complex through the presence of two distinct kinds of specialised (and apparently conducting) tissue. They have been particularly thoroughly studied in *Polytrichum,* where our knowledge of them goes back at least to the fundamental researches of Haberlandt. Collectively these tissues are known as hadrom and leptom (analogues respectively of xylem and phloem) but the constituent elements are conveniently termed hydroids and leptoids (Fig. 20). The wider, thicker-walled hydroids are in the centre, the narrower, thinner-walled leptoids surround them and both are somewhat intermixed with elongate parenchymatous elements. It is easy to see the parallel here with the situation in a simple protostelic fern (sporophyte), and with it the presumptive evidence of conduction. With the recent work of Hébant[172-5] the extent of this structural parallel has been greatly highlighted, for by the use of both optical and electron microscopy he has revealed an impressive series of similarities, both in development and in

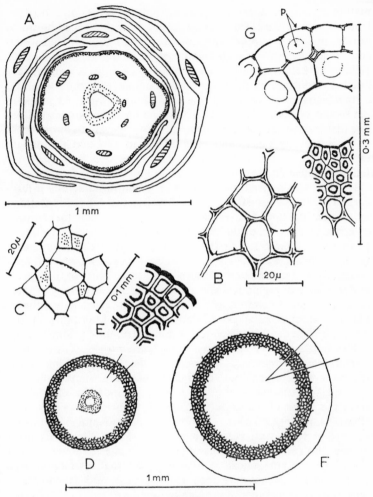

Fig. 20

A. Transverse section of stem of *Polytrichum* sp. showing (somewhat diagrammatically) central hadrom surrounded by leptom, nine 'leaf trace' strands and the bases of five leaves. B. and C. A few cells of hadrom and leptom respectively (after Hébant). D. *Polytrichum* sp., T.S. seta. E. A few cells enlarged, as indicated. Note concentration of thick-walled elements peripherally and presence of central strand. F. T.S. stem of *Sphagnum palustre*, showing position of narrow (often coloured) thick-walled elements. G. Detail of this stem. Note thin-walled cells externally, in three layers composing 'hyalodermis'. p. 'pores' (which are true holes in cell walls).

ucture, between hydroids and tracheids, and between
nd sieve elements. In the former only the special bordered
n the latter only those complete perforations which would
clusters of pits into true sieve areas, appear to be lacking.
himself is so impressed that he seeks a direct evolutionary
etween the two,[173] through the agency of some remote ancestral
form which could have borne this kind of hydroid-leptoid strand in
the tissues of both generations. In this he raises a big question but
one that is outside the present context. More relevant to our purpose
is his interesting suggestion, in another paper,[176] that throughout
moss evolution there has been a steady regression away from the
fully developed conducting strand as we know it in *Polytrichum*.

Lorch[267] had noticed, many years ago, that in numerous pleurocarpous mosses (Isobryales, Hypnobryales) the conducting strand
was better developed in erect subaerial shoots than in prostrate
creeping stems. Dendroid mosses would often display collapsed
hydroids in their creeping stems whilst those in the erect shoots
were healthy. It seems likely that many ectohydric mosses will have
no call for water-conducting tissues in the lower parts of their stems
and it is easy to see how, given a change in the physiology of these
organisms, central strands as a whole, with all their highly differentiated anatomy, could be rendered obsolete. Interesting in this
connection are the facts of central strand distribution among mosses.

We notice that genera such as *Rhacomitrium* (Grimmiales),
Orthotrichum, Ulota and *Cryphaea* in the Isobryales and many
Hypnobryales, which lack central strands, are among the best
examples of ectohydric mosses. Central strands are also absent
from the stems of the quite unrelated *Leucobryum* and *Sphagnum*
(Fig. 20F), in both of which the water economy is manifestly of the
ectohydric type, with the older parts of shoots commonly dead.
Hébant has carried out a general survey[176] and has shown how, on
the whole, the anatomical equipment for internal conduction tends
to be particularly ill-developed in Isobryales and Hypnobryales
whilst, by contrast, it is exceptionally well developed in *Polytrichum*
and its allies. He has divided mosses into six categories which turn
upon the central strand, leaf-trace, and midrib equipment in each
moss. Even in *Polytrichum* the elements in leaf traces coming into
the axis from leaves do not all link up with the central strand, many
vanishing in the cortex (cf. Fig. 20A). In many Eubryales and some
Dicranales hydroids from the leaf midrib scarcely begin to penetrate
even the cortical region of the axis. In others, the stem has a central
strand but the leaf lacks a midrib altogether. To him *Polytrichum*
is primitive, the others in varying degrees derived.

Leaf midribs can display a highly organised level of tissue differentiation, the significance of which is probably not even now fully understood. Among the best examples are species of *Mnium* and *Bryum* (Fig. 21A). In them the transverse section of the leaf midrib reveals four distinct kinds of cell: (1) upper and lower epidermis; (2) a group of relatively large, angular cells in the position occupied by xylem in the leaf of a vascular plant; (3) a compact group of minute, thin-walled cells reminding one of phloem; and (4) as a buttress against the latter, a group of narrow cells with very thick walls. Such a display of tissue differentiation invites comparison with what is seen in the leaves of vascular plants but any apparent resemblances are something of an illusion. In evolutionary origin and in ontogeny the two are completely different. Moreover, we know as yet far too little about the precise function of these tissues to make many generalisations about them. The large, prominent cells of the second group specified above are the 'guide cells' (German, '*Deuter*') of taxonomists. Functionally, they must be the hydroid members, the thin-walled elements that lie below them (*Socii* of some authors) being the leptoid equivalent, in the strand. Any reliable information that we have, however, will have come from the experimental physiologists and it is to their work that we must turn next to discover what is known of the pathways of conduction and translocation in bryophytes.

Regarding water conduction, notwithstanding the pioneer work by Haberlandt, and various studies by others who came after him, many points remained unresolved when Mägdefrau[278] undertook his investigations in the early 1930s. At that time Bowen[46] had just published her conclusion that most water conduction in *Polytrichum* (and some others) was by external capillary channels, but he considered her methods to have been insufficiently critical. Mägdefrau himself established that both external and internal pathways of water conduction are important, external conduction depending largely on the form and arrangement of the leaves and the internal pathway being prevailingly in the central strand. At ordinary temperatures a relative humidity of at least 90% would be necessary for internal conduction alone to be sufficient.

More recently Zacherl[476] used fluorescent dyes to trace the path of internal conduction in selected mosses. In such a genus as *Mnium* he found conduction in the central strand of the stem and in the leaf midrib. The leaf lamina and the ground tissue of the stem were secondarily supplied from these sources. The specialised midrib tissues from the leaf, ending blindly in the stem cortex, and thus forming so-called 'false leaf traces', acted after the manner of wicks and 'drew' the water current across the intervening parenchymatous tissue. In species of *Polytrichum,* where at least part of the leaf

E

trace links up with the central strand of the stem, he found internal conduction was localised throughout in these tissues. Zacherl found that leaves near the apex were supplied first, lower leaves later. In moss stems which lacked a central strand he believed no internal conduction possible.

Bopp and Stehle[43] also used fluorescent dyes for their work on the flow of water in *Funaria hygrometrica,* with special reference to the pathway of supply from gametophyte to sporophyte. They found that both internal and external conduction were involved in the passage up the gametophyte. As regards the supply to the sporophyte, experiment showed that the very narrow space which separated gametophyte central strand from the haustorial cells of the foot received the dye early. Once the fluid was in the sporophyte the central strand of the seta was important for conduction. Removal of the calyptra led to an increased rate of flow. These two examples serve to emphasise the close link that exists between anatomical structure and the physiological processes which such structure makes possible.

Still more recently, attention has turned to the problem of translocation in the leptoids. If hydroids are water-conducting, then we may well look to these more slender, thinner-walled elements to play a role akin to that of phloem in vascular plants. Thus, Eschrich and Steiner[116–17] have shown that C^{14}-labelled assimilates moved in the leptom cylinder of *Polytrichum commune* at a velocity up to 32 cm per hour. They went on to relate the fact of rapid translocation to the histological equipment available. In this work many of their findings coincided with those of Hébant. Like him, they showed the widespread occurrence of callose on plasmodesmata in leptoids, and they confirm that few of the specialised leaf midrib cells succeed in linking up with their counterparts in the central strand of the stem. It is surprising that they should have observed the 'guide cells' (Deuter) joined by a 'leptoid bridge' to the leptom of the stem strand. It seems that even current physiological researches are leaving us a long way from understanding fully the tissue differentiation of mosses.

It must be stressed that bryophytes differ fundamentally from vascular plants in their water relations. They are able to undergo severe drying out, only to recover immediately in the first shower of rain. This, which is an adaptation to intermittent (as opposed to high or low) water supply, is expressed by the term pollacauophyte. Accordingly, any xerophytic adaptations which have been attributed to mosses or liverworts (cf. Watson, W.[450]) will not operate in quite the same manner as they would in higher plants. Repeatedly it has been shown (cf. Clausen[85]) that it is the ability of a particular species to withstand drying out, rather than its ability to hold water as

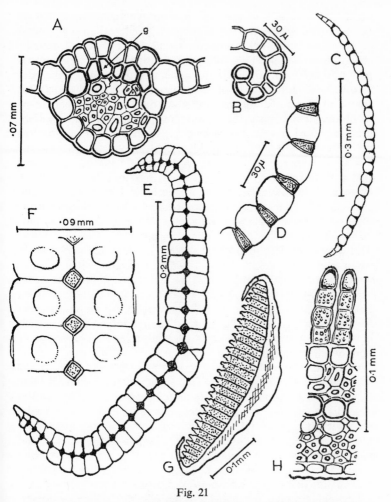

Fig. 21

Moss leaf structure. A. and B. Transverse sections of midrib and margin respectively of leaf of *Bryum pseudotriquetrum* var. *bimum*. g. 'guide cells' (Deuter). Beneath these are seen a small group of thin-walled cells and a massive development of strengthening tissue. C. T.S. leaf of *Sphagnum recurvum*. D. Detail of a few cells. Note chlorophyllose cells (stippled) exposed abaxially. E. T.S. leaf of *Leucobryum glaucum*. F. Detail of a few cells. Note interrupted central layer of chlorophyllose cells (stippled) and hyaline cells with large round 'pores' (holes) in their walls. G. T.S. leaf of *Polytrichum juniperinum* (diagrammatic). Note margins incurved over part of the lamellae. H. Narrow strip of cells enlarged from T.S. leaf of *Polytrichum formosum*, showing two lamellae. Note depth and complexity of midrib tissue on which these lamellae rest.

such, that affords a measure of its xerophytic adaptation. All such devices as the white (dead) hair points of the leaves in some mosses, or the in-curling thallus margins (protected by ventral scales) in some liverworts, must be interpreted with caution; although they may well serve to protect more vulnerable tissues from the effects of direct insolation. It is possible that thickened cell walls (widespread in leaves of both liverworts and mosses) have a part to play in holding tenaciously that minimum quantity of water which is necessary for survival. Lorch believed so, but it is likely that their precise role is not yet fully elucidated.

The resistance of bryophytes to desiccation indeed, is a subject that is only imperfectly understood. It has long been known, for example, from the early investigations of such men as Irmscher[210] and Malta,[281] that different species possess widely divergent powers of resistance. Thus, *Fontinalis squamosa* (an aquatic moss) is said to be unable to withstand a week of air-drying; whilst *Philonotis fontana* died only after fifteen to twenty weeks of this treatment; and the well-known xerophytic moss *Grimmia pulvinata* withstood sixty weeks in a desiccator at 20°C. More recent investigators have been Mirimanoff-Olivet,[290] Höfler,[186] Ochi[310] and a number of others. Hosokawa and Kubota[194] tested many Japanese epiphytic species for ability to survive in different relative humidities. The conclusion must be that genuine differences between species exist but the underlying mechanism eludes us.

On the question of resistance to heat as such Lange[248] has shown a strong correlation between heat resistance and habitat. Thus, *Gymnomitrion obtusum* (moist mountain rocks) and *Plagiothecium curvifolium* (shaded woodland sites) were damaged by 70°C; whilst amongst species capable of surviving temperatures approaching 110°C were *Pleurochaete squarrosa, Tortula ruralis* and *Tortella nitida,* all of them inhabitants of rocks (or other surfaces) commonly exposed to intense insolation. Lange's experimental plants were thoroughly dried over phosphorus pentoxide, then exposed to the given temperatures for half-hour periods. Biebl,[29] more recently, found no marked difference between tropical and temperature bryophytes in respect of resistance to extremes of temperature.

As regards mineral nutrition of bryophytes, the work of Tamm[421] on *Hylocomiun splendens* constituted something of a landmark. This plant produces annual increments in the form of clear-cut, frond-like 'segments' which facilitate comparative analyses. Tamm found that the percentage of nitrogen, phosphorus and potassium decreased with increasing age of the segment whilst calcium increased steadily. He deduced that *Hylocomium splendens,* with its lack of living contact with the substratum, would obtain the bulk of its modest mineral requirement in the form of 'tree leachate', i.e. the washings from leaves of overhanging or adjacent branches of trees. Fallen leaf litter could also be a source but could be detrimental owing to its smothering effect. There was also atmospheric dust washed down by rain. Only the nitrogen supply remained something of a puzzle.

Grubb, Flint and Gregory,[159] who more recently have assessed the mineral nutrition in some common epiphytic mosses, obtained results which accorded only in part with those of Tamm for *Hylocomium*. They believe that much may turn on the power of a particular species to 'redistribute' its supply of a given element (from older to younger parts). They also invoke the possibility of nitrogen-fixing bacteria on leaves playing a part in satisfying the nitrogen requirements of mosses. Streeter[412] found that the content of potassium, sodium, calcium and phosphorus in *Acrocladium cuspidatum* varied with site and season. In a recent, very useful summary this same author[413] underlines the important fact that many different workers on the 'calcicole problem' have concluded that calcium concentration and pH operated as two quite separate factors. He also alludes to work (Hoffman,[184] Voth & Hamner[436]) which has demonstrated the importance, on occasion, of phosphorus and nitrogen levels.

Studies in the respiration and photosynthesis of bryophytes, which currently claim the attention of some physiologists, must be regarded —because of their inherent complexity and the shortage of space— as beyond the scope of a short treatment such as this. The older observers discovered much about the distribution in bryophytes of the cruder and more easily identified products of metabolism, such as starch and oil (cf. Rancken,[350] Lorch,[267] Joensson & Olin[218]) but it is only as new and more refined techniques have become available that many of the associated problems could be attacked at the necessary sophisticated level. Often it is the ecologist who is most directly interested in the physiology of the plants whose distribution puzzles him. Thus, Streeter's ecological summary[413] touches on many of the physiological processes of bryophytes and we shall see more of their relevance in the next chapter.

In the space of a single short chapter it has been possible to touch only lightly on selected topics in this very wide field. Even this however may have sufficed to show how closely the three—morphogenesis, anatomy and physiology—are interconnected. The resources of all three are combined in the quest to see the moss or liverwort as a developing, functional whole. We may conclude by referring briefly to some studies which, whilst essentially anatomical or biochemical, nevertheless have an important implication for systematics. The first of these is the subject of 'oil bodies'.

These microscopic deposits, which occur in the cells of many liverworts, received some attention in an earlier chapter (cf. p. 44). They are quite unconnected with deposits of fatty oil (which are a widespread food reserve in bryophytes, as in higher plants). They vary from the massive deposits which occur in certain cells of some thalloid liverworts (e.g. *Lunularia*) to collections of minute granular bodies (up to ten to twenty in a cell) which are found in the leaves of many Lejeuneaceae. The smaller type of 'oil body' in fact occurs in a great number of Jungermanniales. Schuster and Hattori[387] have provided a comprehensive and fully illustrated treatment of them in the Lejeuneaceae. A further detailed and wide-ranging account comes from Inoue.[208] The emphasis throughout is on their taxonomic significance, for the precise form of oil body can be a reliable specific character; or it may be invoked at generic or family level. Meanwhile the link between these structures and the physiology of the plant as a whole remains obscure.

In recent years phytochemists have begun to turn their attention towards bryophytes. Their enquiries can be expected to yield results of interest both to physiologists and to taxonomists. Already we have the broad survey by Huneck[198] and scattered papers by others. Among them we find McClure and Miller[274] assessing flavonoids in mosses, and Lewis[259] surveying the sugar alcohols in bryophytes and bringing chemical evidence to support the new taxonomic position of *Plagiochila carringtonii* (long placed in *Jamesoniella,* until perianths were found). The future field must be enormous. The approach is biochemical but the end that is served is systematic. Similar ends are currently being served by some kinds of anatomical work. Good examples are the large-scale survey by Bischler[32] of the 'systematic anatomy' of the stem in Lejeuneaceae and the recent examination, by Kawai,[229] of midrib structure in the leaves of mosses. There are also some studies of ultra-fine structure, with either scanning or transmission electron microscopes, that hold an obvious taxonomic significance. A good example is the electron-microscopic study of spore wall structure in Musci by McClymont and Larson.[275] Researches in all these areas are likely to extend greatly in the years to come.

I O

ECOLOGY

Bryophyte ecology is the study of the part played by mosses and liverworts in various plant communities. It is concerned with the importance of particular species and with their relationship one to another. The subject has grown with increasing rapidity in recent years. Many nineteenth-century bryologists who knew well the habitats of the plants which they studied must indeed have been skilled in some facets of the subject. They would have known the habitat preferences of different species and the groups of species which it was usual to find growing together. Most often, however, they were not experimentalists, and it is the experimental side of bryophyte ecology which has extended so greatly during the past forty years or so.

The species list from a particular habitat may well form the starting point, and provided it be carefully compiled, and the habitat clearly defined, this list can be of value. Bryophyte lists in general works such as *The British Islands and their Vegetation*[422] have tended to come from diverse sources and to be uneven in quality. Those in this great classic of plant ecology came in part from W. Watson, himself a pioneer in this field, whose numerous papers on the bryophytes and lichens of various habitats[451-3] formed the foundation on which much subsequent work was to be built. Even this able bryologist did not concentrate on the analysis of particular communities. Rather did he compile lists of the bryophytes of a type of country, making a compilation from diverse sources and localities. Nevertheless, these early papers are a mine of useful information and can still be consulted with profit.

The species list tells us at a glance the contrast between the component bryophytes of acid heath and those of chalk down, or again between the colonists of bare peat and the pioneers on maritime sand. From such lists we learn the value of 'indicator' species, such as *Ditrichum flexicaule* and *Encalypta streptocarpa,* which point at once to base-rich conditions; or the species of *Polytrichum and Rhacomitrium* which are almost without exception acid 'indicators'. Again, there are those species with habitat so circumscribed that their presence on a list can reveal the precise conditions prevailing. *Leskea polycarpa,* on tree roots liable to flood-water, is an example. Or the species individually may mean little, but taken together they may define a habitat with precision. Thus, the presence together of the liverworts *Nardia scalaris* and *Diplophyllum albicans,* and the mosses *Ditrichum heteromallum, Oligotrichum hercynicum, Pohlia elongata* and *Polytrichum urnigerum* will be highly suggestive of a fine-grained mountain detritus, moderately acid, on a fairly steep slope and at a not excessive altitude. A short list from a particular niche is often more informative than a longer, comprehensive one. Even so, at its best the species list is only a beginning.

A natural step forward is to include a consideration of succession. Ecology owes this term to the pioneer American ecologist F. E. Clements,[86] and it alludes to the orderly sequence of change in vegetation as it proceeds towards what Clements called the natural 'climatic climax' where some kind of stability or equilibrium is reached. One of the first successional sequences (termed seres) in which the part played by bryophytes was clearly set forth was that on burnt heath (Fritsch and Salisbury[127]). Here these and subsequent observers have found that *Funaria hygrometrica* is among the first colonists, to be followed in sequence by *Ceratodon purpureus* and then *Polytrichum juniperinum* and *P. piliferum* in association with lichens of the genus *Cladonia.* The first two phases are passed through in perhaps a couple of seasons, but the third phase commonly lasts for many years. The phases are not always clear-cut but the abundance of *Funaria* has usually been held to be linked with high pH and high nutrient status (especially potash) which follow the fire. In a recent study, however, Hoffman[185] has questioned this.

Again, Richards[359] investigated bryophyte succession on sand dunes and concluded that at Blakeney mosses were prominent at early and late stages but were of less importance during the middle phase of succession. Thus, species such as *Bryum pendulum* and *Brachythecium albicans* can be early colonists in the lee of the main seaward dune ridge, whilst far back on the fully fixed dune pasture quite another group of species, including *Rhytidiadelphus triquetrus* and other large 'pleurocarps', will achieve prominence. Richards noted some of the special adaptations possessed by the widespread dune moss, *Tortula ruraliformis,* and Willis[466] has taken the matter further. Gimingham[142] found that *Bryum pendulum* and *Barbula fallax* had special powers of withstanding periodic burial by sand, growing up through the layers deposited. More stable conditions prevail where the big pleurocarpous species achieve their maximum. Quite different is the shingle succession described by Scott.[391]

Leach,[254] studying non-calcareous screes, found bryophytes prominent in three distinct types of habitat, on the surfaces of rocks, in chinks between them, and thirdly as part of the invading 'heath' flora. Once more a succession may be detected as one passes from the second of these stages to the third. The liverwort *Diplophyllum albicans* Leach found to be the principal colonist of the chinks, whilst the later stages were marked by an increasingly varied bryophyte element. On the Cader Idris massif in North Wales, on acid granophyre block scree, I have found the small moss *Grimmia doniana* prominent on the surface of the blocks, *Rhacomitrium lanuginosum* the principal colonist in the deep clefts between blocks, and about a dozen species of bryophyte commonly contributing to the more stable vegetation that marked the ultimate phase in the succession. This included tall species of erect habit, like *Dicranum scoparium* and *Polytrichum alpinum,* species with rich and varied branch systems like *Hypnum cupressiforme* and *Rhytidiadelphus loreus,* and several robust leafy liverworts such as *Plagiochila spinulosa* and *Anastrepta orcadensis.*

Heath, dune, scree are but three out of the countless illustrations that could be taken of successional sequences. So too, with bryophyte ecology as a whole. It is a very large subject and space will admit of our discussing only a few selected topics. The following will now be considered briefly: (1) attempts to recognise and define bryophyte communities; (2) growth forms; (3) quantitative methods; (4) the time scale in succession; (5) refinements of habitat study; (6) single species (autecological) studies. Finally, we shall conclude by examining a few more general matters that are relevant to bryophyte ecology as a whole.

(1) *Concept of community.* Clements put forward the notion of a plant community as a close-knit unit, with a life of its own which he likened to that of a living organism, divisible into phases of growth, maturity and senescence. There were those who asked to what extent bryophytes could form a community of this kind. Or was the bryophyte cover little more than the haphazard juxtaposition of species, often in keen competition but unworthy of the name of community at all? One or two examples may help to clarify the

position. On old elders (*Sambucus nigra*) one may find a mossy covering containing perhaps the following eight species: *Orthotrichum affine, Zygodon viridissimus, Bryum capillare, Ceratodon purpureus*, Brachythecium rutabulum, B. velutinum, Hypnum cupressiforme* and *Cryphaea heteromalla*; and with them, close-pressed to the bark, may be the liverworts *Frullania dilatata* and *Metzgeria furcata*. On a number of grounds this may be said to constitute a bryophyte community. On old elders, over a wide geographical area, this group of species, or one very like it, will be found; the species will be competing with, and reacting upon, one another, but mutual benefit is also possible through conservation of rainwater, ameliora-tion of microclimate and so on; the different species, each with its own morphology and growth rate, are in a sense parts of one whole; and finally, at a certain point in the life of the elder tree the community had a beginning, it achieves maturity when all species are present and fully grown, and in the foreseeable future, some time after the death of the tree, the life of the community will end. Some case can be made then for the recognition of bryophyte communities. Difficulties arise, however, when one considers the manner in which a particular bryophyte cover may change over the years (see p. 143).

Herzog,[181] in an enumeration of the bryophyte communities of the Black Forest mountainous region, has designated them by reference to their most characteristic species. He sometimes finds a single species that defines the community, as with *Polytrichum strictum* (*P. alpestre*) which indicates drying out of the upper layers of bogs. It is described as pushing up through the *Sphagnum* cushions and in its establishment it is associated with certain lichens (*Cladonia, Cetraria*), but with few bryophytes. We can see it doing the same thing in some mountainous parts of Britain. More frequently Herzog's communities are marked by at least two characteristic species, as in the *Philonotis seriata-Bryum schleicheri latifolium* community, where associated species are numerous and together give a clear picture of the prevailing conditions. The list includes *Pohlia albicans* var. *glacialis, Bryum duvalii* (*B. weigelii*), *Calliergon* (*Acrocladium*) *stramineum, Dicranella squarrosa* and several large species of *Scapania*. It resembles what I have myself seen in some of the high flushes in the Cairngorms. The area with which Herzog dealt, together with other parts of southwest Germany, has recently been the subject of an exhaustive treatment by Düll.[115]

The followers of the Zürich-Montpellier school of phytosociology

---

* *Ceratodon purpureus*. cf. Barkman for the conclusion that this species is most characteristic on willows (*Salix* spp.).

(founded by Braun-Blanquet over thirty years ago) insist on a strictly hierarchical system of nomenclature when describing communities. Such a system, whilst fundamental to Braun-Blanquet's procedure, can lead to an almost too rigid circumscription of the different entities. So far as bryophytes are concerned, we see it in Gams's early article[135] in the *Manual of Bryology* and again in the sequel[136] to this which he wrote twenty years later. Barkman,[20] too, has employed such a system in the descriptive part of his monumental '*Phytosociology and ecology of Cryptogamic epiphytes*'. Less rigid accounts of bryophyte ecology, but nevertheless with some concept of community well to the fore, are those of Doignon[110] (Fontainebleau), Gaume[139] (Brittany), Paton[321] (Wealden sandstone of Kent) and a host of others. Indeed, all lists from defined habitats hint at community structure. If the concept is to be real, however, then the component species might be expected to present characteristic growth forms in different cases. It is this aspect that we shall discuss next.

(2) *Growth forms.* The emphasis was decidedly on growth form when Gimingham and Robertson[144] published their '*Preliminary investigations on the structure of bryophyte communities*'. These authors built on earlier foundations laid by Meusel[289], and others before him. Their first working scheme of growth forms in bryophytes was subsequently modified (Gimingham and Birse[143]), although its main essentials were retained. Five categories of growth form are recognised—cushions, turfs, canopy formers, mats and wefts (Fig. 22). Most of these are further subdivided, but the canopy formers (with such mosses as *Thamnium, Climacium* (Fig.22J) and *Mnium undulatum*) compose a single group on their own. A size difference separates large cushions (such as *Leucobryum*) from small cushions (such as species of *Grimmia* (Fig.22A) and *Ulota*). Again, tall, short and open turfs are recognised, with *Polytrichum commune, Bryum argenteum* (Fig. 22B) and *Polytrichum aloides* respectively as an example of each. *Breutelia chrysocoma* exemplifies a special kind of tall turf with abundant rhizoidal development on the stems; *Campylium stellatum* and *Sphagnum* (Fig.22D) are cited as illustrations of 'tall turfs with divergent branches of limited growth'. Mats prove difficult and in the four subdivisions are included such diverse forms as *Eurhynchium striatum* (rough mat), *Frullania tamarisci* (smooth mat), *Eurhynchium praelongum* (thread-like) and finally members of both Marchantiales and Metzgeriales under the heading of thalloid mats. Wefts are more straightforward, and although in the original version the authors separated 'spreading branched' from 'pinnately branched' wefts they have later replaced this subdivision by one based on frequency of rhizoids. Hence *Hylocomium splendens*

(Fig. 22H) is a relatively rhizoid-free weft, *Thuidium tamariscinum* one with frequent tufts of rhizoids.

The need for some workable system of growth-form classification had indeed long been felt and on page 128 of his classic *Bryogéographie de la Suisse* Amann[6] set forth in outline a system put forward two years earlier by Herzog. This made the primary division into solitary (*Rhodobryum roseum*) and colonial forms, and the latter group was further subdivided into eight categories, several of which were the same as those of Gimingham and Robertson. The system was less comprehensive than theirs and is now of mainly historical interest. Gimingham and Robertson explained that a single species could at different times fall into different growth-form categories. Even so, the decision as to where to place a particular example will not always be easy to make. Fortunately, Gimingham and his co-workers have followed the original paper with numerous studies in which the scheme has been employed in the ecological analysis of bryophyte vegetation. The system has also been used by various workers in other countries, notably in America and Japan.

Iwatsuki[212] is a good example of a Japanese ecologist who has used this tool, but in doing so he has made a number of slight modifications of the system of Gimingham and Robertson. The most notable is the erection of a special category, 'pendulous forms' for such subtropical genera as *Barbella, Pseudobarbella* and *Floribundaria*. Iwatsuki's long paper, incidentally, is an immensely thorough survey of epiphytic bryophyte communities in Japan, and one that is germane to much else that appears in this chapter. It would repay close study. This author has since published, in collaboration with Hattori,[213–14] a long series of additional papers on cryptogamic epiphytes in Japan.

The principal conclusion which Gimingham and Birse[143] and Birse[30,31] have reached from subsequent work has concerned the

---

Fig. 22

Some growth forms of mosses. A_1. Outline of 'small cushion', in diagrammatic vertical section, as seen in *Grimmia pulvinata*. A_2. A shoot from the same. B. 'Short turf', as seen in *Bryum argenteum* (simplified to emphasise parallel arrangement of shoots). C. Single shoot of the same. D. and E. General plan and detail of *Sphagnum recurvum*, a special kind of 'tall turf', with fascicled branches of limited growth. F. Part of a 'mat' of *Hypnum cupressiforme* var. *filiforme*, in which numerous parallel branches adhere closely to bark. G. Detail of one branch. H. *Hylocomium splendens*, 'weft' showing four years' increments, each with bipinnate branching. J. *Climacium dendroides*, single erect shoot of dendroid form which, with others, makes the plant a 'canopy former'. K. Detail of one branch.

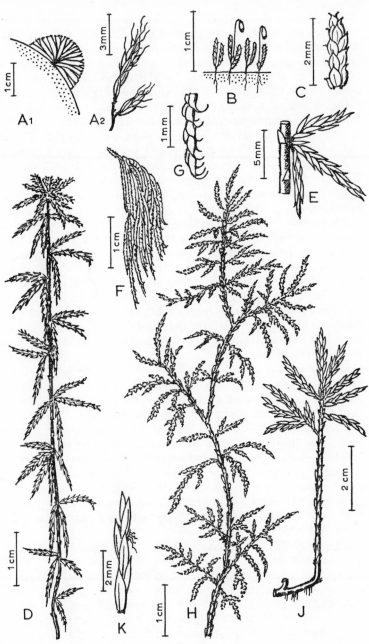

Fig. 22

fitness of particular growth forms for particular habitats. These authors point out that 'the features of community structure may reappear under recurring habitat conditions irrespective of the species present'. The terms in which this 'community structure' is expressed are the growth forms. An illustration may be given. The authors[143] are concerned with the zonation of bryophytes in a deep ravine. From running water one passes successively through four zones: *Eurhynchium riparioides, Cratoneuron filicinum, Conocephalum conicum* and finally *Thamnium alopecurum*. The presence of these last two is used to point the habitat requirements of 'thalloid mats' and 'canopy formers'—i.e. no submersion but high relative humidity. The question that remains unanswered is why other species, exhibiting these same growth forms, do not appear in the positions indicated.

The growth-form system of Gimingham and his co-workers is an important contribution to bryophyte ecology; but it is essential that we should understand both its advantages and its limitations. It rests on externally visible structural pattern alone, and therein lies its weakness from the interpretative viewpoint. We may recall that Raunkiaer's life form classification of higher plants hung on the position of over-wintering buds. This expressed a real biological difference between one life form and another, thus imparting a fuller meaning to the system than any comparable arrangement of bryophyte growth forms can have. Thus, in Gimingham's system the two genera *Sphagnum* and *Polytrichum,* which must be very different biologically, are placed together under the general head of turfs. Again, in an examination of sand dunes, the point is made by Birse that a 'tall turf with erect branches' is characteristic of the drier areas, a conclusion reached because the single species *Tortula ruraliformis* tends to prevail on such sites. The fact that it was a 'tall turf' must surely be ranked subordinate to those physiological characteristics which Willis[466] was able to reveal as contributory to its survival in this rigorous habitat. Recognition of different growth forms has its uses but biologically it must be less significant than Buch's[57] separation of bryophytes into endohydric, ectohydric and mixohydric groups (cf. Chapter 9).

(3) *Quantitative methods.* As a relatively early example of quantitative methods we may refer to studies of the bryophytes of chalk grassland. Here, Hope-Simpson[190] laid the foundation by preparing lists of species which occurred respectively in mature ('closed') grassland and in seral stages. Some fifty sites were sampled and each species was allotted a status, 'abundant', 'frequent', 'occasional' or 'rare'. As a result he was able to name the eight species which were most constant in this habitat. What we do not know is the size of the

fifty areas surveyed nor the means whereby each species was allotted its status. Later Watson[446] studied a more limited area, using a definite and readily repeatable sampling method. He agreed with Hope-Simpson in placing *Pseudoscleropodium purum* as the most generally abundant moss of chalk grassland, but a numerical approach demonstrated that it was approximately three times as abundant as its nearest rival.

Perring[328-9] used a still more clear-cut sampling method to establish, among other things, that in north Dorset the mosses *Weissia microstoma, Eurhynchium swartzii, Neckera complanata* and *Fissidens cristatus* were confined to the steeper slopes of aspects between south-east and west. On the other hand, *Hylocomium splendens* and *Rhytidiadelphus squarrosus* were limited in Dorset to slopes between north and east, whilst from comparable chalklands near Rouen they were absent altogether. In the past ten years quantitative methods have impinged ever more insistently and there have been numerous examples of bryophyte ecological studies in which ordination techniques have been employed. Papers by Bunce,[59] Foote,[125] Proctor[340-1] and Yarranton[474-5] all illustrate this trend. Such methods are particularly valuable where complex data have to be analysed and conclusions drawn.

(4) *Time scale in succession.* Some examples of bryophyte succession were considered on an earlier page; but we did not probe deeply the question of the time scale of events. This is always of interest, and certainly varies greatly. Thus, some fallow field communities can achieve completion within a single season, i.e. in a matter of months after the 'bare soil' becomes available for colonisation. Elsewhere the situation may be very different. If one is following the course of events on a fallen bough, through the long period of decortication and decay, until a patch of raw humus is all that remains of the branch that once supported a group of epiphytic mosses, then surely one must speak of a succession of communities. There will be, first of all, the community of epiphytes borne by the living branch. This will be followed by a group characteristic of dead wood in a very different microclimate (almost that of the forest floor). In due course the decorticated bough or log will present a surface of soft, well-decayed wood for colonisation by such species as the liverwort *Nowellia curvifolia,* the loss of bark surely heralding the onset of a new community again. When at last the branch crumbles to raw humus a final community ensues, with *Campylopus piriformis* probably the chief member. Doignon[110] has reported that this sequence of events may take some twenty-five years in the forest of Fontainebleau. In northern Scandinavia, according to Barkman,[20] it may well take several hundred. Doignon[111] also states that when fire

destroys old heather moor (*Callunetum*) with its attendant bryophyte cover of such species as *Pleurozium schreberi, Hypnum cupressiforme* var. *ericetorum* and *Dicranum scoparium,* it may well be thirty years before this type of bryophyte flora is restored, so gradual are the various stages of regeneration.

Barkman[20] gives a vivid account of succession on a living tree, which was studied by Doignon in the forest of Fontainebleau. Lichens colonised first and twenty-five years of the tree's life had passed before the liverwort *Frullania dilatata* arrived, closely followed by species of *Orthotrichum, Ulota* and other genera. When the tree was some thirty-five years old there came mosses of the genera *Anomodon* and *Neckera,* and the liverwort *Porella platyphylla.* Finally, after forty years, *Leucodon sciuroides* and *Zygodon viridissimus* colonised the mature tree.

The study by Bliss and Linn[36] of 'old-field' succession in the Piedmont of North Carolina is an example of a rapidly changing environment supporting a succession of bryophyte communities. The expression 'old field' is in general use in America for fields that have been allowed to revert as cultivation marched westward, and every ecologist familiar with the American scene knows the meaning of 'old-field pine' (*Pinus* spp.) and 'old-field cedar' (*Juniperus* spp.) as the names for important conifers which step in after the land has been abandoned. Bliss and Linn found the richest bryophyte community that of the first full season after cultivation had ceased, some seventeen species succeeding the pioneer, *Sphaerocarpos texanus.* At the four- to seven-year period pine became conspicuous, and from eight to fifteen years saw the transition to full dominance of pine. Fifteen years after abandonment the number of species of bryophytes had dropped to five. Thereafter changes were rather gradual over many years. Observations suggested that after a lapse of some period well over 120 years pine would be superseded by deciduous trees; and these would entail an entirely new range of bryophyte communities.

Apart from successional sequences of this kind, there are instances where a favoured site has been known to support the same group of bryophytes, or maybe a particular species of moss or liverwort, for 80 to 100 years, or as far back as records go. Yet we may know little of its fluctuations in abundance over the years, of how often and how successfully it reproduces, vegetatively or by spores, or what is the exact age of the larger cushions or tufts. Some moss shoots reveal their age readily and reference has been made (Chapter 9) to the 'segments' of *Hylocomium splendens,* each representing one annual increment. Again, one sometimes finds the moss *Breutelia chrysocoma* with several successive seasons' capsules at different levels on the old stems, and it may be easy then to assess the age of the tuft. Most often the life of an individual plant or colony is short and lack of change in the site as a whole is something of an illusion.

I am familiar with an outcrop of greenstone on the Isle of May, Firth of Forth, where the cushions of the rather rare moss *Grimmia stirtoni* have been known for at least eighty years. By marking a permanent quadrat on the outcrop one was able to see that the life of a particular cushion may be short indeed, and within seventeen months there had been big changes in the distribution of the colonies of this moss.[447] The chart quadrat method (where each plant is drawn or symbolised) may be usefully employed in work of this kind.

The relationship between bryophyte pioneers and subsequent vascular plant colonists can also be studied by this method. Although many people think of bryophytes as early colonists (often following algae and simple lichens) which give place to vascular plants later, this is only partially true. On many sites, on boulders, roofs, cliff-faces and some kinds of unstable scree, only casual vascular plants may succeed in gaining a footing; and the bryophyte cover may be a kind of environmentally conditioned 'micro-climax' of its own. A certain stage in development is reached. Then some natural phenomenon, such as frost action or snow melt, results in the removal not only of the little community but of part of the sub-stratum too. A fresh surface is laid bare, and the whole process starts all over again. In mountain cliff habitats some such cycle of events must be very common. One would like to know more of the exact time-sequences involved.

(5) *Refinements of habitat study*. Life on the trunks and branches of trees will provide a good example of modern emphasis upon the more refined study of the habitat; for here mosses and liverworts are an important component of the flora, and Barkman[20] has given a penetrating analysis of the whole environment. He points out that different sides of a tree-trunk may present very different micro-climates; and this for three main reasons. These concern (a) the aspect of trunk (or branch) in relation to the daily march of the sun, (b) aspect in relation to the direction of the prevailing wind, and (c) the inclination of the trunk from the vertical position. Thus, deep in a ravine, a tree may have the same epiphytic vegetation all round the trunk. Barkman cites Kraemer[240] as having been the first to point out the advantages of a leaning trunk. The value of the off-vertical habitat so created seems to lie mainly in its power to catch and retain atmospheric precipitation. Once a bryophyte cover has been formed, this enhances the effect and cuts off the

under surface of the leaning trunk from any appreciable water supply.

No full grasp of the varied habitat conditions offered to epiphytes is possible without some appreciation of the differing potentialities of the bark of different species of tree. Apart from properties associated with bark character (rough or smooth, hard or soft, and so forth) two chemical factors, as Barkman shows, are very important. These concern nutrient supply and acidity. The rich epiphytic flora associated with elder (*Sambucus nigra*) is partly related to the high nutrient status of its bark. This has 8–12% ash content in the dry bark and in reaction it approaches neutral (pH range 5·5 to 7·0). By contrast, species of *Picea* (spruce), *Abies* (silver fir) and *Pinus* (pine) tend all to be low in nutrient status and markedly acid in reaction. On these conifers the bryophytes are few. Trace elements such as boron, cobalt and molybdenum, which in recent decades have proved important in crop nutrition, are not known to exert any influence in bark ecology.

There is a much richer epiphytic flora in woods in western Britain, as compared with the south-east. The cushions of *Leucobryum glaucum,* ordinarily a ground species, may be seen ascending the trunks of trees; the communities are richer in species, and the total bryophyte cover on favourably placed trees is much more extensive. Barkman[20] refers to *Dicranum fuscescens,* which, he says, will grow on tree bark in wet forests, on rotten logs in moist ones and on the ground (raw humus) in dry ones. Thus, he considers that it behaves as a microclimatic indicator. In the Rothiemurchus pine forest of Inverness-shire it can be seen in all these three situations, in different parts of the forest.

Maritime conditions, as might be expected, affect the epiphytic flora of trees that grow in coastal dunes or other situations within reach of salt spray. According to Barkman, the high humidity induced by such a situation (coupled with the high water capacity of elder bark) enables mosses like *Acrocladium cuspidatum* and *Leptodictyum riparium* to appear on sand dune elders; for these are not common epiphytes.

Richards[360] referred to certain mosses that were able to grow in the splash zone and could thus be classed as genuine halophytes. The outstanding example is *Grimmia maritima,* its blackish-green cushions being a feature of so many maritime rocks in western Britain. *Ulota phyllantha* at times grows with it, not far above high-water mark. Shacklette[393] studied a halophytic community of bryophytes on Latouche Island, Alaska, and found *Tortella flavo-virens, Pottia heimii, Camptothecium sericeum, Hypnum cupressi-forme* and *Grimmia maritima* all to be salt-tolerant. The special conditions of this type of habitat, and the way in which they operate on the few bryophytes capable of colonising such ground, await further investigation.

Shacklette in the same paper discussed the peculiar group of bryophytes restricted to areas that are exceptionally rich in copper, a subject to which Persson[332] had earlier given attention. Persson gave figures showing the copper content of the mosses themselves (in parts per million) and the pH of the substratum. He did this for species of 'copper moss' belonging to the three genera *Mielichhoferia, Dryptodon* and *Merceya,* and although the copper content varied it tended always to be high and could be as high as 675 parts per million. On Latouche Island Shacklette detected several distinct cuprophile communities, and in one case the liverwort *Gymnocolea acutiloba* grew practically pure over a wide area, forming a kind of 'liverwort peat' several inches thick. He also named a group of bryophytes tolerant of high gypsum content, and another of species which grew on substrata rich in sulphide of iron. These, however, are not peculiar to such habitats in the way that the copper mosses are, for they are seen to include species well known in quite different habitats in Britain. Among the gypsum-tolerant are *Dicranella cerviculata* and *Cephalozia bicuspidata,* widespread on peat in this country, and among the sulphide-tolerant are *Rhacomitrium fasciculare, Oligotrichum hercynicum* and *Nardia scalaris,* which abound on acid rock and scree habitats generally. A more up-to-date statement on copper mosses was published by Shacklette[394] in 1967.

The investigations of Grønlie[156] have shown that a special vegetation develops on ground exceptionally enriched by the droppings of sea birds (guano). He worked on the famous bird cliffs of Røst, in the Lofoten Islands of northern Norway, and (following earlier workers) he termed the plants tolerant of such conditions ornithocoprophilous. Grønlie listed over forty bryophytes on the bird cliffs of Røst, and named a number of them as strongly ornithocoprophilous. Among these were *Ceratodon purpureus, Eurhynchium praelongum* and *Mnium hornum,* all of which are also prominent on the Isle of May (Watson[444]), where great colonies of sea birds nest. It may be that in the Arctic there are no other habitats which supply the high nutrient status that these species require. On the other hand, on the Isle of May there are 'basicole' species elsewhere on the island which find no place in the heart of the bird colony, where a special combination of high nitrogen status and high disturbance level prevails.

When modern methods are employed some progress can be made towards solving the old problem (cf. Richards[360]), i.e. isolating the effects of a particular ecological factor. This is what Clausen[85] did in her admirable study *Hepatics and humidity*. She worked on an area in East Jutland sufficiently limited to enable her to make a thorough study of the hepatic flora and to correlate distribution with accurate measurements of relative humidity. So refined were her methods that they allowed extreme localisation of readings; and Clausen quotes an example from the steep bank of a lane where *Nardia scalaris* enjoyed a relative humidity of 85% whilst only 5cm away, on a slight prominence, *Frullania tamarisci* was subject to 55%. It is likely that in the future an even finer analysis of micro-habitats will be made.

(6) *Autecological studies*. It was by such a study that Tamm[421] was able to advance our knowledge of the moss *Hylocomium splendens,* from the ecological as well as the physiological stand-point (as noted in Chapter 9). The work of Tallis[419–20] on *Rhacomitrium lanuginosum* is another example. It has thrown light on the ecology of an outstandingly interesting and important moss. For here is a plant which occurs in such abundance on certain kinds of mountain-top detritus that it has given the name 'Rhacomitrium heath' to the vegetation which it dominates. Yet one sees it also (as already mentioned) as the pioneer on some kinds of block scree; or again as a late-comer in the succession on certain types of old bog. I have seen it in such a role on the Isle of Barra, Outer Hebrides, where it formed mound-like cushions half a metre across. It abounds on many types of substratum in base-poor mountain country; it is surprisingly widespread on limestone in northern England, and, most remarkable of all, occurs in a few places on the chalk of the South Downs, growing in the turf quite close to known calcicole species. Such a moss presented a challenge indeed, so that the comprehensive study by Tallis was all the more welcome. More-over, as Tallis himself points out, *Rhacomitrium lanuginosum* is a moss of world-wide distribution.

In his first paper Tallis assessed the role of *Rhacomitrium lanu-ginosum* in different communities. In his second paper he considered growth form, growth rate, manner of reproduction and several features of the plant's physiology. Among other things, he showed that the white hair-points on the leaves were longest in dry, sunny weather in the summer; also, it was between May and August that maximum growth occurred and lateral branches were best developed. The annual growth increment was always small, however, varying from 5 to 15mm in length. This slow growth rate, and the fact that some, at least, of its physiological activities are at their optimum within a narrow temperature range (13–15°C), suggested that this moss might succumb easily in the face of competition. Tallis was inclined to attribute part of its success in many barren and inhospitable sites to a capacity (for which he finds some evidence) to render soluble some normally insoluble components of the substratum.

More recent examples of the autecological approach are Forman's work on *Tetraphis pellucida*,[126] Hoffman's study of *Funaria hygrometrica*[185] and Briggs's thorough examination of the ecology of several species of *Dicranum*[48] (which he was also studying from the taxonomic angle). One could mention, too, the stimulating work of Longton and Greene[264] on *Polytrichum alpestre* in the Antarctic. In every case our knowledge of the intimate biological relationships of a species is thereby increased and we acquire a fuller picture of the 'niche' it fills. Sometimes such knowledge will help workers in other fields. Some years ago H. Watson[449] showed how useful a knowledge of the ecology of conspicuous bryophytes could be to the practical forester. On occasion a moss will have an interesting 'story to tell', as in the case of *Mnium undulatum*, the presence of which (Hörmann, *in litt.*) led to the rediscovery of a lost village in a forest in Austria. This moss had selected the locally enriched ground where the former village had existed before its total destruction during the Thirty Years' War. All such examples, however, highlight the individual species.

We have examined six topics, in all too little depth. Of course there are many others that call for attention, but space will allow only the briefest reference to a few of them. The student must understand that bryophyte ecology is a subject of almost limitless scope and there are many parts of it in which our knowledge is still very rudimentary. This is true, for instance, of most aspects of tropical bryophyte ecology, although the 'epiphyllae', and their relationship to the leaves on which they grow, have been receiving attention in several parts of the world (cf. Chen & Wu,[78] Mizutani,[296] Winkler[470]). Argument still rages as to what methodology is best and Lye[272] is surely right to claim (in connection with community structure) that 'no system of classification can be regarded as correct, or even better than another system'. In his own work on oceanic bryophyte communities in Norway he has combined astutely the quantitative and qualitative approaches. The student who would range widely in the available literature will find some familiarity with the plants themselves is an almost essential pre-requisite. Only then will he be able to appreciate the immense interest of an outstanding contribution such as Ratcliffe's '*Ecological account of Atlantic bryophytes in the British Isles*'.[352]

Again, it is only when one knows some of the more important species of *Sphagnum* that one can read with interest and profit of the differing parts they play in bog ecology. This in itself is an enormous branch of the present subject, and one with a vast literature. Accounts by Rose,[368] Ratcliffe & Walker,[353] Boatman,[37] Gimingham,[145] Tallis,[420] Svensson[415] and many others deal with different aspects of it, whilst Clymo[87,88] has done much to explore the more strictly physiological processes involved. Among other topics, he has been concerned with the differential rates of breakdown of various species.

Finally, we cannot close this chapter without making brief reference to the role of bryophytes in assessments of atmospheric pollution. Their significance in this context is second only to that of lichens. Often the two groups of organisms are considered together. In the last few years interest in the whole subject of pollution has accelerated greatly. Hence the papers by Rao and Leblanc[351] (in Canada) and Gilbert[141] (in Britain) are particularly timely. Both were led to the same conclusion, that sulphur dioxide was the most formidable pollutant. In his thorough examination of the Newcastle area Gilbert has brought rigorous methods to bear on a problem that has long bothered bryologists in many parts of the world. His reference to a great city as a 'bryophyte desert' is no exaggeration.

GEOGRAPHICAL DISTRIBUTION; GEOLOGICAL
HISTORY; CYTOGENETICS AND SPECIATION

In this chapter we are concerned with three big, interrelated topics. The distribution of families, genera and species in the world today must to some extent reflect the history of these groups. Palaeobotany can tell us more of this. Moreover, the cytogeneticist can provide us with an additional point of view; and, more important, he can tell us something of how new species arise and how existing ones are interrelated. A survey of bryophytes would be very incomplete without some enquiry into these three branches of knowledge.

When we see a particular moss or liverwort growing it is of interest to enquire in what other parts of the world it grows. Again, when we examine the 'annotated list' for some area or country, it is instructive to split it into its components, viz. groups of species displaying similar distribution patterns in the world as a whole. From such enquiries springs the science of phytogeography, and the information can be set forth for species, genera or whole families. What is revealed will reflect events which took place in the remote past. At the same time it may tell us something about the mechanisms of speciation and dispersal. To be useful our information must be based upon a sound taxonomy, and it must be as complete as possible.

Bryophytes, at all taxonomic levels, tend to show wider distributions than flowering plants. Many families are found throughout the world. So too are many of the larger genera, such as *Polytrichum, Weissia* (sensu lato), *Grimmia, Bryum* and *Brachythecium* among mosses; *Plagiochila, Lophocolea, Radula* and *Frullania* among liverworts. Even at the species level the list of cosmopolitan species is quite large. Some, like *Ceratodon purpureus, Tortula muralis* and *Funaria hygrometrica,* must surely owe part of their world-wide distribution to the activities of man, for they are rather in the nature of 'international weed species'. Others, however, among cosmopolitan mosses do not grow on waste ground, wall tops or cultivated soil, for example *Gymnostomum calcareum* and *Drepanocladus uncinatus. Hypnum cupressiforme,* often regarded as the most 'variable' British moss, is said to be cosmopolitan, but Herzog[180] has aptly remarked that if we were to accord specific rank to three of the most marked 'varieties', *lacunosum, resupinatum* and *ericetorum,* we should see in the first a Mediterranean, in the other two an Atlantic, type of distribution.

It is indeed to Herzog's monumental *Geographie der Moose* (1926)[180] that we owe much of the foundations on which others have built so freely in the years that have followed. It may be that he was a shade too greatly influenced by certain ideas that were prevalent in his day, for instance a rigid concept of phytogeographical regions and the inherent unlikelihood that two forms found in far-separated regions of the world would belong to a single species. Inevitably, too, he must have been handicapped by some faulty taxonomy and by extensive gaps in existing knowledge of the bryophyte flora of many countries. But, these considerations apart, he managed a remarkably complete and penetrating analysis which can still be read with profit today. In bringing together information amassed by systematists who went before him, he showed how bryophytes could display a great variety of localised distribution patterns. To this end he reviewed them, family by family. From Herzog's work it was evident that certain 'patterns' of distribution tended to recur, again and again. So one would come to recognise groups of families and genera that displayed circumboreal, Mediterranean, pan-tropical, bipolar and other kinds of distribution. To examine one's own flora in this way is to see it in a new light. Allorge[4] (much later) did this for the French and Spanish 'Pays basque'; Gaume[138] did the same for Brittany. Each examined the situation at species level, at the same time recognising the existence of certain prevailing types of distribution pattern. Gaume had a long list of circumboreal species, shorter lists for a dozen other categories. Especially interesting, perhaps, were species common to Atlantic Europe and Atlantic North America, like *Andreaea rothii* and *Campylopus flexuosus*; or common to the Mediterranean region and Western North America, like *Tortella nitida, Funaria attenuata* and others. It is always the patterns which show gaps, or disjunctions, that are of the greatest interest.

Quite recently there have been further important papers that have dealt with these and related types of distribution. The links between the bryophyte floras of Eastern Asia and North America have been examined, for instance, by Schofield[378] and by Iwatsuki & Sharp.[215] Others, like Duda & Váná[114] for Czechoslovakian liverworts, have made extensive use of maps to display the details of distribution in different species, at the same time drawing attention to phytogeographical implications. Furthermore, an enormous amount of information has been streaming in concerning the detailed composition of the bryophyte flora of many lands (cf. Schultze-Motel,[380] van Zanten,[432] both on New Guinea, Florschütz[124] on Surinam, Pócs[336] on Vietnam, Hong[189] on S. Korea, Reed & Robinson[354] on Thailand, Bizot[34] on Cuba, Robinson[365] on Colombia, to mention only a selection), with the result that the plant geographer has constantly to revise 'his facts'. New taxonomic studies, too, can force him to do this. He is in an altogether stronger position than he was forty years ago.

Because this is so, it is particularly valuable to have had, for liverworts, the exceedingly important contributions of Fulford,[133-4] and still more recently, of Schuster.[384,386] Both have dealt especially with liverworts of the Southern Hemisphere and through their work many of the salient questions have had fresh light brought to bear on them. Both have been concerned, for example, with the South American–Australasian and other disjunctions, with endemism, dispersal mechanisms and evolutionary questions. Clearly our interpretation of the facts of geographical distribution must depend greatly on our views regarding the possibility (or frequency) of long-range dispersal.

Herzog[180] was quite convinced that, despite the lightness of spores, long-range 'jumps' were almost as rare among bryophytes as among flowering plants. Others have thought differently and Proctor, V. W.,[342] was not the first to invoke the agency of migratory birds in long-range dispersal when he claimed that waterfowl might assist in the spread of a species of the aquatic liverwort genus, *Riella*. True, a species sometimes turns up in a new place (in a well worked country like Britain) fifty to a hundred miles from its nearest known station, and spores seem to be the obvious means of spread; but this is not a really long-range move. Were the latter at all frequent it is unlikely we should see the many striking (and surely meaningful) distribution patterns that claim our attention. Fulford[130] certainly has supported Herzog and has pointed out that the short duration of viability of many liverwort spores would effectively prevent their travelling great distances. However, it would be rash to claim that the long-range 'jump'—involving thousands of miles— never occurred. Moreover, many bryophytes which rarely or never produce spores, manage to spread effectively. Often, of course, it is difficult to rule out the possibility of chance introductions by man. The subject is beset with difficulties.

Fulford[130] drew attention to the bipolar type of disjunct distribution shown by the leafy liverwort genera *Blepharostoma* and *Anastrepta,* among others. She found several genera, all of which displayed 'many primitive or simplified characters', having a widely disjunct distribution in the Southern Hemisphere as a whole. Schuster[386], in a very recent paper, has taken this theme further. Basing his conclusions on his own extensive researches in the Antipodes and sub-antarctic, he finds some genera (e.g. *Archaeophylla, Phyllothallia,* etc.) with a range that is highly disjunct yet restricted to the far south. Another group (including *Hymenophytum, Acromastigum, Zoopsis,* etc.) is predominantly sub-antarctic but shows limited penetration of the Northern Hemisphere. Both authors agree about the preponderance of primitive forms involved. Schuster calls them ancient survivals, including under this head a formidable list of endemic genera, and finds the evidence strong enough to deduce the origin of most Jungermanniae in 'Panantarctica'. He is convinced that in the past the liverworts of the far southern regions have been insufficiently studied and grossly misunderstood. Unanimity on taxonomic matters within this group, however, can scarcely be said to exist as yet, so that phytogeographical details will turn to some extent upon the taxonomic views of the author who cites them.

As regards mosses, Herzog's account[180] contained many fine examples of disjunct distributions, at generic level, most of which must still be valid today. Two of these are the curious large-spored genus *Gigaspermum,* with representatives in Australia, South Africa and Morocco; and *Echinodium,* which is found in the Azores and Madeira, reappearing only in many parts of Australasia. He also drew attention to whole families, such as Neckeraceae, Hookeriaceae and Sematophyllaceae, which are predominantly tropical but have a few outlying species spreading up into the north temperate zone. He found just four species in Europe out of the 660 known in the family Hookeriaceae. The same kind of situation prevails at the generic level, for example in *Breutelia.* This genus, however, also tends to show an Atlantic type of distribution. Hence our only W. European species, *B. chrysocoma*, belongs to the group recently analysed in some detail by Ratcliffe[352] for the British Isles and even more fully by Størmer[411] for Norway.

Endemics, whether at family, genus or species level, have always held a special interest. They are of two principal kinds. On the one hand there are forms so recently evolved that they have lacked the time to achieve a wider distribution. On the other hand there are those so ancient that, having vanished from their former stations elsewhere, they have become restricted to a single country. Still others may have evolved quite far back in time but because of geographical barriers remained endemic. Island floras abound in such examples. Thus, the moss flora of Tristan da Cunha[109] has fifty-nine endemics out of 128 species. Endemic genera will be far fewer. Herzog cited *Rutenbergia* from Madagascar and numerous others. Many genera, and even whole families, are restricted, or largely restricted, to a single broad region of the world but just escape the definition 'endemic'. Most eloquent among plants (or groups) of restricted distribution, for the phytogeographer, are those which are clearly recognisable, on morphological grounds, as ancient forms. They may be true endemics, like many liverwort genera cited by Fulford and by Schuster, or near-endemics like the moss *Pleurophascum grandiglobum* which stands alone taxonomically, is confined to Tasmania and New Zealand, and was described by Herzog as a 'living fossil'; or they may merely be genera like *Haplomitrium* among liverworts and *Buxbaumia* among mosses, which display a widely scattered pattern of disjunct distribution. To conclude this short treatment of a very big topic, it may be apposite to quote the cautious note struck by Schuster:[386] 'Present evidence does not often allow us to distinguish between ancient overland dispersal, and more recent dispersal by spores'. Even this warning, however, need not blind us to the fact that a certain kind of endemism and instances of disjunct distribution afford the best evidence that phytogeography can supply in support of the ancient origin of particular forms. For more than that we must consult the fossil record.

It has sometimes been said that bryophytes are so poorly represented in the fossil record that a discussion of these fossils can contribute nothing of value to the general question of moss and liverwort evolution. There might have been some truth in this view fifty years ago; but it is far from true today, and is likely to become less so as time goes on. True, in certain groups, e.g. Anthocerotae, Andreaeidae, on which information would be particularly desirable, no early representatives are known. What is encouraging however is that even in the last ten years the number of fossil bryophytes has increased significantly (cf. Lacey[243]).

We are fortunate in having two long accounts of the fossil contribution, both published within the last fifteen years. The first was the *Geological Annals of the Bryophyta,* by Savicz-Ljubitzkaya and Abramov,[374] the second being the splendidly illustrated account by Jovet-Ast[225] in Boureau's *Traité de Paléobotanique*. There is also, still more recently, the useful review by Lacey. All these authors agree that, so far as pre-Tertiary fossil bryophytes are concerned, real strides have been made only in the last fifty years, when the newer palaeobotanical techniques had become available. Nineteenth-century accounts tended to lack the necessary microscopic detail. It is because of such lack of detailed information that palaeobotanists have been forced to place many of their finds in comprehensive 'form genera', such as *Muscites* and *Hepaticites*; a practice which makes no claim more specific than that a 'fossil moss' or 'fossil liverwort' is at hand. Where the evidence warrants it, Schuster would like to employ a name that tells us more of relationships—hence his introduction of *Treubites* for Walton's fossil liverwort that showed a manifest link with *Treubia*. All too often, however, such affinities are less apparent and for the time being the form genus has to stand.

Savicz-Ljubitskaya and Abramov[374] gave figures for the total numbers of known pre-Tertiary fossil bryophytes. Their figures for the Mesozoic—fourteen liverworts and three mosses—have since increased sharply, to thirty-four liverworts and eight mosses. Lacey[243] makes this point; he also observes that Palaeozoic liverworts are up from seven to nine, mosses from nine to seventeen, if one compares the Russians' account with that given by Jovet-Ast[225] less than ten years later. Tertiary fossil bryophytes are much more numerous, but are of less interest because they tend to be so much like known living forms. Few though they may be as yet, the early fossil bryophytes, as we shall see, can provide us with some answer to two important questions. The first is the question how far back in time the two great groups, mosses and liverworts, were differentiated. The second (which all keen students of living bryophytes will want to ask) is the question: did there exist, a hundred or more million years ago, in early Mesozoic or Palaeozoic time, bryophytes that were quite unlike any present-day living form?

The year 1925 was something of a landmark, for it saw the publication of the first account of Carboniferous liverworts in sufficient detail to leave no doubt as to what kind of plants they were. On this occasion,[439] and again in 1928,[440] Walton made it clear that there existed in the Coal Measures certain liverworts which bore a fairly close resemblance to some modern Metzgeriales. Thus, *Hepaticites willsi* was thalloid after the manner of *Riccardia*; but *Hepaticites kidstoni* was a clearly leafy form in which the existence of appendages of two distinct sizes (lateral leaves and dorsal scales) suggested an affinity with *Treubia* (Fig. 23A–C). Such an affinity for this fossil has never been seriously contested and it duly finds its place (as *Treubites kidstoni*) in the modern monograph of Treubiaceae by Schuster and Scott.[388] A third species, *Hepaticites lobatus,* has prompted Schuster[384] to erect the new genus *Blasiites* for it. If Walton's fossils established the antiquity of the order Metzgeriales,

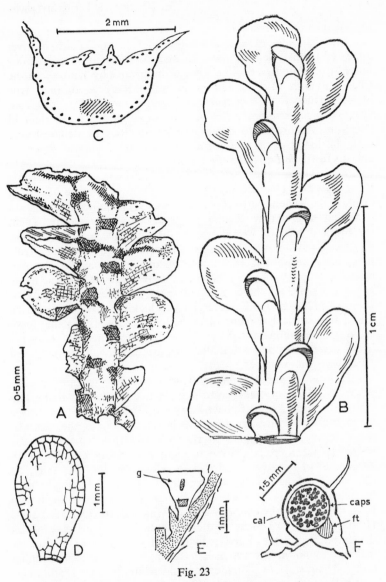

Fig. 23

A. *hepaticites* (*Treubites*) *kidstoni* (after Walton). B. *Treubia tasmanica*, small part of shoot, drawn from plants grown in greenhouse at Reading, the leaves more distant than usual. C. Transverse section, showing massive axis and parts of insertions of large and small leaves. Dark dots indicate peripherally placed oil-body cells. Hatched area: ill-defined central strand. D–F. *Naiadita*. D. leaf, E. 'gemma cup' (g), F. sporophyte (after Jovet-Ast's adaptations of the original drawings by Harris). cal. calyptra; caps. capsule; ft. foot.

the same cannot be said for other liverwort orders. No member of the Marchantiales, Jungermanniales or Sphaerocarpales has yet appeared from rocks of Palaeozoic age. Jovet-Ast[225] refers to some tantalisingly '*Anthoceros*-like' spores discovered by Knox[237] in the Coal Measures of Fife, but proof was lacking as to their true identity. Hueber's[196] *Hepaticites devonicus* (from the Upper Devonian), as Lacey reminds us, is the earliest certainly known thalloid liverwort. According to Schuster[384] its affinities lie demonstrably with the living genus *Pallavicinia* and he has re-named this very ancient fossil *Pallaviciniites devonicus*.

A fossil which stands very much alone is the remarkable Mesozoic plant *Naiadita*. This had long been a well-known feature of certain layers in the Rhaetic when Harris,[164] after a very detailed examination involving several new techniques, revealed its true nature. Then, for the first time, it was shown to display, time and again amid its exceedingly abundant remains, a whole series of bryophyte features. Axis and foliar structures, rhizoids, gemmae, archegonia, even a sporophyte of general form strongly reminiscent of the living genus *Corsinia,* all these were laid bare in remarkable clarity of detail (Fig. 23D–E). Only one organ was lacking; nowhere could be found unequivocal remains of antheridia. The work on this fossil stands alone among fossil bryophyte investigations, not only for its great wealth of detail, but also for the strange, baffling picture of a liverwort that emerges.

In a point-by-point summary Harris was able to bring a considerable weight of evidence to suggest that its nearest living ally may have been *Riella* (Sphaerocarpales). The form of the leaves (though not their arrangement, which was moss-like), the rhizoids, the relation of archegonia to perianth and the structure of the spores (Fig. 15A, p. 90), were brought reasonably into line with *Riella,* but in the sum of all its characters *Naiadita* may fairly be claimed as unique. Given such detailed knowledge of a single fossil bryophyte, bryologists would easily be tempted to do too much with it, were it not for the cautious note on which Harris ends his paper. He insists that we beware of the temptation to regard *Naiadita* as the common ancestor of diverse living forms just because it combines features found in certain genera belonging to different families today. The history of individual organs, he reminds us, is still dark. Not all have heeded this warning. Schuster, for example, in his recent book,[384] sees the possibility of deriving both Marchantiales and Sphaerocarpales from a *Naiadita*-like form. In this connection we must not forget that all the evidence indicates that *Naiadita* was an aquatic liverwort. Aquatics are notoriously apt to have become modified in unusual ways. Meantime, this fossil, at once so abundant and so isolated, must remain something of an e

In recent years some light has been shed on the remote pas
the Marchantiales. Lundblad[271] and Townrow[427] have describe
fossils from the Rhaetic–Liassic of Scania and the Middle Triassic
of Natal respectively (See also Harris[165]). The upshot is that we
know that there existed in Mesozoic times members of the order
Marchantiales that must have been extraordinarily like some living
genera of that order. In *Ricciopsis scanica* Lundblad describes and
figures a fossil liverwort which bears a convincing resemblance to
some modern species of the genus *Riccia*. In *Hepaticites cyathodoides*
Townrow sees resemblances to *Cyathodium* in habit, and in the
structure of midrib, rhizoids and ventral scales, but the 'pores'
that occurred on the upper surface of the thallus were unlike those
of the modern genus.

The examples quoted above are but two of the several known
Marchantiales from the Mesozoic; and we have seen Sphaerocarpales represented by *Naiadita*. Of the remaining orders of liverworts,
however, there is no fossil of comparable age. A fine series of Jungermanniales from Tertiary times has been found preserved in the
Baltic amber, but we have no knowledge of the group from either the
Palaeozoic or the Mesozoic period. Furthermore, on the whole
subject of the early history of sporophytes the fossil record is (with
occasional exceptions such as *Naiadita*) completely silent. On
gametophyte evidence, however, the Metzgeriales stand established
as the Order of liverworts of maximum antiquity. It could be argued
that the known fossils are still too few for us to lean heavily upon
this conclusion. *Sporogonites,* a curious Devonian fossil sporophyte, tentatively attributed to the bryophytes by Andrews,[15] seems
too imperfectly understood to help us very much.

For mosses the fossil record is equally patchy. Indeed, as regards
really early forms it is considerably more incomplete than that of
liverworts. The earliest examples, stratigraphically, appear to be
Muscites polytrichaceus, and *M. Bertrandi* figured by Walton[440]*
from the Carboniferous. Apart from that the record from the
Palaeozoic would look bare indeed were it not for the sensational
series of beautifully preserved fossil mosses described by Neuburg[304–6]
in recent years from Permian beds in Angaraland, USSR. Thanks
to their good state of preservation and the use of modern peel-
transfer techniques, a wealth of detail is made available sufficient
to assign one group tentatively to the Sphagnidae, another to the
Bryidae; although it must be admitted that the lack of any information about the sporophytes is a serious shortcoming.

To contain the first group consisting of the three genera, *Junjagia,*

* Described earlier by Renault & Leiller and by Lignier respectively.

l *Protosphagnum,* Neuburg erected the Order
tween them these genera shed a most interesting
istory of the Sphagnidae. Arnold[18] had already
um itself went back at least to the Cretaceous;
ed both leaves and spores of this genus from the
r Nuremberg, in Bavaria; and Jovet-Ast[225] lists
es of *Sphagnum* (or *Sphagnumsporites*) from the
e based largely upon fossil spores. What Neuburg
has ...... nething both older and decidedly different from
Sphagnum as we know it today. In varying degree these fossil leaves
resemble *Sphagnum* in displaying a 'cell network' composed of two
distinct types of element arranged in a regular and characteristic
pattern; but the pattern is not quite that of the modern *Sphagnum*
leaf. The extent of differentiation was less complete, and the hyaline
units appear themselves to have been divided on occasion into
'triads' (Fig. 24D,E). Furthermore, as Jovet-Ast points out, the broadly oval leaves of *Vorcutannularia* were arranged in striking rosettes
up to 3 cm wide, and reminiscent of *Bryum giganteum* (Fig. 24B).
Finally, the leaves in these genera differ sharply from modern
Sphagnum in that they have a midrib and sometimes traces of
lateral veins as well (Fig. 24B,D).

In Neuburg's second group of Permian mosses is a group of
genera which have to be classed merely as 'Eubryales Incertae
sedis'—true mosses whose precise systematic position is uncertain.
The genus *Intia,* for example, with four species, bears some remarkable
resemblances to the modern genera *Mnium* and *Bryum.* The leaf
dimensions and shape, cell structure and thickened, toothed border
(Fig. 24A,C)—all are there. Among the other genera can be seen a
range of features (e.g. the branched lateral veins of *Polyssaievia*)
which amply separate them from what we know in modern moss
gametophytes. The absence of all trace of reproductive organs or
sporophyte, however, makes further conjecture rather fruitless.

Muscites guescelini, described by Townrow[427] from the Triassic of
Natal, may well be an example comparable with the above, for it
could not be brought perfectly into line with any known living moss.
It lacked the arresting features of most of Neuburg's fossil genera,
however, and in some respects came close to modern mosses of
the family Leucodontaceae. Again one regrets the absence of
reproductive organs.

The early fossils, then, can furnish some answer to the two
questions posed on an earlier page. Firstly, some groups at least
of both mosses and liverworts occurred in highly differentiated
form as long ago as the late Palaeozoic. Fossil thalli and leafy
shoots proclaim this to be so, though we know nothing of the

sporophytes of these plants. Secondly, *Naiadita,* together with many of Neuburg's moss genera, tell us that there did exist, in the Mesozoic and the Palaeozoic respectively, bryophytes that were quite unlike those we know today.

Of the considerable wealth of more recent fossils we have said nothing. This is because they have little to contribute to the particular questions that we have raised. In quite another field, that of Quarternary botany, or the study of changing floras during and since the Ice Age, bryophytes are becoming increasingly important. Associations of arctic species found in interglacial or very early post-glacial deposits in parts of southern England testify to the conditions then prevailing. The recent literature has many scattered records of 'sub-fossil' finds of this type. Among the more notable

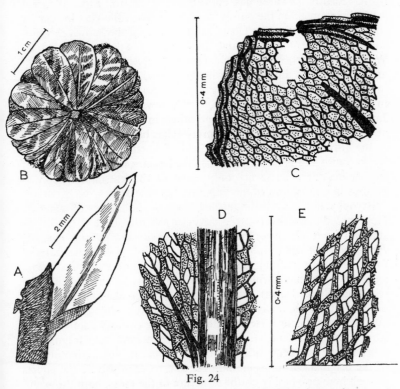

Fig. 24

Fossil mosses of Permian age. A. *Intia vermicularis,* leaf and part of stem. B. *Vorcutannularia plicata,* rosette of leaves. C. *Intia variabilis.* Part of leaf, showing border, nerve and leaf cell structure. D. and E. *Protosphagnum nervatum.* Portions of leaf, D. near nerve, C. away from nerve, to show 'pattern' of cell arrangement. (All after Neuburg)

are papers by Godwin & Richards,[147] by Landwehr[246] and several by Dickson,[102] who for many years has made a special study of this subject. He has recently given us a full statement on the British moss flora of the Weichselian Glacial period (from *c.* 50,000 to *c.* 12,000 years ago). As Godwin himself has indicated elsewhere,[146] these records are of particular interest when they reveal that subarctic or arctic mosses, such as *Paludella squarrosa, Meesia triquetra* and *Acrocladium trifarium,* formerly grew in regions whence they have long since gone. Such facts give a new dimension to our thoughts on plant geography. Udar* has helped in another direction, by clarifying the place of bryophyte spore studies in palynology.

Turning now to our third topic, we find that Allen[2,3] twice reviewed the extensive literature on the genetics of bryophytes. More recently Lewis[260] has contributed a general paper of great intrinsic interest. On the purely cytological side, following the earlier survey by Sinoir,[399] comprehensive catalogues of the known chromosome numbers of mosses and liverworts have been provided by Wylie[471] and Berrie[27] respectively. Lowry,[269-70] Anderson,[8-9] Bryan[52-4] and Steere[410] are among those in the United States who have made important contributions in recent years. So too with Vaarama[429] in Finland; Yano,[472-3] Tatuno,[424-5] Segawa[392] and Inoue[207-9] in Japan; Mahabele,[280] Khanna[231] and others in India; Lazarenko[251-253] and Visotska[434] in the USSR. Ramsay[349] has begun to make headway in the hitherto almost untouched field of Australian moss cytology, whilst Smith and Newton have published a most important survey of British material. These are only some of the outstanding contributions in a large and expanding 'area' of research.

Bryophytes have a distinguished place in the history of genetics. Not only did the liverwort *Sphaerocarpos* provide a classical instance of sex chromosomes, but mosses of the genera *Funaria, Physcomitrium* and *Bryum* furnished Wettstein[457-9] with the material for experiments of fundamental importance, more than thirty years ago. He made successful inter-generic crosses; he showed that the resulting plants leaned more towards the female than the male parent in their characters and was thus led to postulate a maternal cytoplasmic influence. He repeatedly raised new leafy gametophytes from protonema that had grown as a regenerant from fragmented seta; and he was hence able to obtain plants carrying high multiples of the original chromosome number and to demonstrate the effects of such high polyploidy on the structure of the moss, both gross and minute. Finally, Wettstein and Straub[460] succeeded in synthesising a 'new species' of the genus *Bryum, B. corrensii,* with twice the nor-

* Udar, R. (1964). Palynology of bryophytes. 79–100 in Nair, P.KK., Advances in Palynology, Lucknow.

mal number of chromosomes, and fully monoecious, unlike *Bryum caespiticium* whence it came. Historically, it is also of interest to note (as Anderson has reminded us) that Wilson[467] in 1908 gave the first correct chromosome count for a moss (*Mnium hornum, n*=6) and at about the same time the Marchals were producing the first polyploids of mosses experimentally. In the 1920s Heitz initiated technical procedures which did much to facilitate subsequent progress.

Since the appearance of Berrie's useful survey of the chromosome numbers of liverworts, much additional information has accumulated and Schuster[384] has provided us with a valuable 'summary statement'. Berrie claimed that the Hepaticae were 'well-known cytologically', but this is only in a comparative sense and it is salutary to be reminded by Schuster[384] of our ignorance regarding such supposedly primitive groups as the entire 'Ptilidiine–Herbertine sequence of families', in the Jungermanniales. We have as yet, indeed, only a smattering of information but even that is sufficient for some tentative conclusions to be drawn. Chromosome number is the first and most obvious class of information one seeks from the cytologist and this is cited as the count obtainable from gametophyte tissue. Comparison of available counts will then tell us immediately if there are instances of polyploidy and it may on occasion give us a hint of evolutionary events on a much wider scale. Chromosome morphology, including the occurrence of odd-sized (often minute) chromosomes with exceptional staining properties, and chromosome behaviour are much more difficult and specialised topics. They too have contributions to make but it is perhaps rash at this stage to attempt to draw far-reaching phylogenetic conclusions from them. We may begin by noting some prevalent chromosome numbers.

In some 75% of all liverwort species investigated so far the basic haploid number is 9. A long series of species of *Riccardia* (and the non-green *Cryptothallus*) are exceptional in that n=10. The situation n=8 occurs in *Riccia, Pallavicinia* and throughout the bulk of the sub-orders Radulinae and Porellinae of the Jungermanniales. Whilst the apparently primitive Calobryales fall into line cytologically with n=9, the very isolated class Anthocerotae reflect their isolation by presenting a unique karyotype, n=5 or 6. It is of great interest that the recently discovered *Takakia lepidozioides* should reveal the lowest number of all—only four chromosomes in gametophyte tissues.[425] Schuster has drawn attention to the fact that n=4 appears also to be the lowest basic number in the green Algae (Chlorophyta), occurring for example in *Ulothrix*. Even this brief summary of purely numerical information is sufficient to suggest strongly that most liverworts have sprung in the first instance from ancestral forms that displayed basic chromosome numbers of 4 or 5. One should however

F*

beware of drawing major phylogenetic conclusions from this evidence alone. Most observers nevertheless would agree that the existence of different basic numbers in major groups is revealing; and this may operate at family level, as Schuster has suggested for the cytological differences between Riccardiaceae and Metzgeriaceae. The picture is however greatly complicated by the structural diversity encountered within each chromosome complement; so much so that when contemporary cytologists (cf. Tatuno,[425] Segawa[392]) make a statement regarding the chromosome complement of a liverwort they express it in terms of the total number of V-shaped and J-shaped chromosomes respectively, together with any differentially staining (heteropycnotic) chromosomes of large (H) or small (h) size. Sex chromosomes (themselves usually heteropycnotic) can further complicate the position. Usually, in these cases (cf. Lewis,[260] Schuster[384]) a simple X–Y mechanism prevails. In some Frullaniaceae, however, the situation is more complex (cf. Lorbeer,[266] Tatuno[425]).

If liverworts present only a small range of basic haploid chromosome numbers and invite phylogenetic speculation, what of the mosses? In Wylie's useful survey of 1957 only 364 species were listed. Five years later Anderson[8] was able to increase this figure to 647. Every year new 'counts' come in from many countries, yet even now it is likely that less than 5% of the world's moss flora are known cytologically. Smith and Newton's massive achievement[402] in making over 800 counts from a total of 225 species and eleven varieties from Great Britain and Ireland serves to stress the contrast that exists between our knowledge of north temperate and our ignorance of tropical mosses from the cytological standpoint. At the same time, the informed commentary of these authors enables us to see, perhaps more clearly than ever before, the pitfalls that have to be avoided in assessing the nature and value of the cytological contribution. In many ways the position is more complex than in liverworts, with an apparently wide range of basic haploid numbers and many surprising inconsistencies within particular groups.

Special interest attaches to the lowest haploid numbers met with, for it is from these that all others must ultimately have been derived. Thus, one can quote $n = 11$, 12 or 13 as representing a predominant karyotype in the orders Dicranales, Pottiales, Grimmiales (and some others); and $n = 10$ in *Bryum*, $n = 11$ in *Pohlia* and so on. By so doing, one establishes the important generalisation that a particular number commonly runs through a long series of related plants; but the fuller meaning of these numbers becomes apparent only in relation to the existence, in certain mosses of $n = 5$, 6 and 7, and in relation to the incidence of polyploidy. These low numbers in fact are uncommon;

we find *Polytrichum* and allies with $n=7$, *Mnium, Orthotrichum, Breutelia* and a few other genera with a basic haploid number of 6, *Rhytidiadelphus loreus* and *Pleurozium schreberi* with $n=5$. Smith and Newton are surely right in concluding that most mosses are 'apparently gametophytically polyploid'.

Inconsistencies abound and the original papers should be consulted for details. We find, for example, the genus *Tortula* with numbers ranging from $6+1\text{m}$ to 60, whilst the adjacent Trichostomeae almost all show $n=13$. In *Sphagnum,* whilst most *'Acutifolia'* and other slender species show $n=19+2$, and most *'Inophloea'* and *'Subsecunda'* (more robust plants) show $n=38+4$, the slender *Sphagnum fimbriatum* is out of step and reveals the diploid figure. *Dicranum fuscescens* appears to be extremely variable cytologically, *D. scoparium* surprisingly constant (at $n=12$). Some inconsistencies, and some cases of apparent aneuploidy, may be attributable to the widespread occurrence of 'm' chromosomes of minute size. In sum, the mosses display much greater diversity than the liverworts, and Smith and Newton suggest that their cytological plasticity may have played a significant part in the comparative success they have enjoyed.

There are many examples of both haploid and diploid figures occurring within a single genus. Sometimes the haploid is a dioecious species, whilst related diploids are monoecious. The best-known example is the genus *Mnium,* where $n=6$ obtains consistently in the dioecious species, those with $n=12$ being monoecious (cf. Lowry,[269] Lewis[260]). Often, however, two figures are found within a single species. Among British examples we find *Pohlia cruda,* $n=11$, 22; *Leptodictyum riparium,* $n=20$, 40, *Drepanocladus revolvens,* $n=10+1\text{m}$, 20 and so on. Lewis cites several exotic mosses in which the example with the higher chromosome number is distinguished by its monoecism, but this does not invariably hold. Sometimes we find evidence of a genuine polyploid series. *Atrichum* is a much studied example (cf. Noguchi and Osada[308]; Lazarenko[253]), with $n=7$, 14, 21; *Distichium,* likewise, reveals $n=14$, 28, 42; and within the species *Physcomitrium pyriforme* gametophytic chromosome complements of 9, 27 and 36 have all been reported. Lewis[260] has alluded to Wettstein's[459] interesting discovery that, in experimentally produced polyploids in this moss, the addition of each successive complement of chromosomes was marked by a significant increase in the volume of the leaf cells. However, with the passage of time these effects were moderated and, in general, there is no corroboration of this condition in naturally occurring polyploids. Smith and Newton,[402] however, refer to the large size of the gametophyte plants in *some* of the examples of *Tortula muralis* with $n=52$. From this fact they deduce that these are recent polyploids, with the size discrepancy not yet

⌐y time; and that, perhaps, in this species polyploidy has ⌐ore than once. Highly relevant to this eventuality is the ⌐r of mosses to regenerate diploid leafy plants from damaged ⌐orophytes (cf. pp. 123–4), a power which—especially in certain types of habitat—could have contributed to the 'cytological plasticity' mentioned above. It is a power which appears not to be shared by liverworts.

Smith and Newton refer to 'diploid' *Tortula muralis* showing $n=52$ (and so it is in relation to haploid plants with $n=26$) but when examined in relation to the lowest chromosome number known in the genus ($n=6$) this figure appears as a high polyploid; and a similar argument can be employed in other genera too. The result is that most numbers seen in living mosses have to be interpreted as in varying degree derivative from lower antecedents. The above authors are emphatic, however, about the undesirability of attempting to draw sweeping phylogenetic conclusions from the numerical pattern emerging. On the other hand, both chromosome number and chromosome morphology (or behaviour) can often tell us when some genus or species has been misplaced taxonomically. Thus, Smith and Newton find that *Amphidium lapponicum* and *Ptychomitrium polyphyllum* both belong (on cytological grounds) in the Grimmiales, not in the Isobryales; and the chromosomes of *Tetraphis* confirm the systematic position it formerly had, near the Polytrichales. Ramsay,[349] finding $n=7$ in *Dawsonia,* brings cytological evidence to support a relationship between this mainly Australasian genus and the very widespread *Polytrichum*. The unusual basic haploid number of *Sphagnum* ($n=19+2$) confirms the extreme isolation of the genus. *Pseudoscleropodium purum* ($n=5$) is found to be cytologically closer to *Brachythecium* (where it was formerly placed) than to members of the subfamily Entodonteae. In such ways can cytology help to resolve doubts regarding relationships.

At a different level it is useful as a tool in the study of some critical species or group. An admirable example is the work of Anderson and Bryan[10] on *Fissidens cristatus* and the closely related *F. adianthoides*. Using material from Nova Scotia, they demonstrated haploid numbers of $12+1$ for *F. cristatus* and 24 for *F. adianthoides* but adduced further cytological evidence in support of the clear separation of these two taxa. In Great Britain the position appears to be less clear, but this does not detract from the intrinsic value of this type of investigation in which cytological and biometric procedures are combined to good effect. A comparable illustration is provided by the earlier work of Müller[299] on the critical separation of *Targionia hypophylla* and *T. lorbeeriana*. All cell measurements (including spore size) run larger in the latter species which is a plant of localised Atlantic-western Mediterranean distribution. The haploid chromosome number is 27, against 9 in the very much more widely distributed and smaller *T. hypophylla.*

Schuster[384] calculates that some 15% of liverworts are polyploid. In any event, the condition is much less widespread than in mosses. Any figures given however will turn on just what constitutes evidence of polyploidy. Among the most clear-cut instances are those where two or more chromosome numbers are recorded for closely related liverworts (as in *Targionia,* cf. also the many instances from mosses quoted above). Schuster quotes examples from the genera *Tritomaria, Chiloscyphus* and others. In by no means all such cases is specific recognition warranted. Paton and Newton[325] found that the 'species' *Pellia borealis* (originally erected for the diploid entity) could not be maintained since the morphological characters said to distinguish it could be found equally in haploid material. Thus we have to see *Pellia epiphylla* as existing at two levels of ploidy. Schuster,[384] rather similarly, has cast doubts on *Calypogeia suecica.* Within *Dumortiera hirsuta* Tatuno[424] has revealed three levels, with the hexaploid the most widely distributed in Japan.

All cytological findings are potentially of interest in relation to both speciation and distribution problems. Moreover, they have to be interpreted in terms of the 'breeding systems' prevailing in the plants concerned. We may usefully conclude this section with a few words on the significance of these systems. Thus, the fact of dioecism carries with it special evolutionary potential, with the possibility of unlimited variation, given the existence of a population of suitable size and diversity. This may be within the limits of the species definition, or beyond them; for there are plenty of records of hybrid sporophytes in the literature, at least of mosses (cf. Andrews,[14] Pettet,[334] Reese and Lemmon[356] for examples at intergeneric level). A high incidence of sterility within such hybrid sporophytes will however severely limit their significance. Dioecism, at the same time, carries disadvantages, the most obvious (as Schuster has pointed out) being the remote chance that a distantly dispersed spore of a dioecious species has of establishing a new outbreeding colony. Schuster[384] remarks that 80% of all liverworts are dioecious and comments in this connection on the abundance, among liverworts, of purely *phenotypic* modifications. Lewis[260] emphasises a different point when he writes that because of the limited range of gametic dispersal, a potential mating group at something like species level is broken into smaller units which are unable to exchange genetic material. In short, this leads him to expect many *genotypic* races in bryophytes. Hatcher[168] has warned us of the frequent impossibility of distinguishing (by traditional methods of examining material) between the genotypic and the phenotypic variant. Lewis is doubtless right to emphasise that population cytology is as yet not a well explored field, especially among bryophytes. All agree, however, in emphasising the 'price' the plant pays for the greater 'security' conveyed by the monoecious state. Since all the gametes from any one gametophyte will be genetically identical, then, barring the possibility of some exceptional event, monoecious taxa have entered a 'phylogenetic blind alley'. A cycle of 'unchanging homozygosity' has set in.

This is equally true of those taxa which have come to rely exclusively on vegetative propagation. Only clonal material is dispersed. Yet there is another side to this argument, for, as Schuster has pointed out, there is a certain 'selection pressure in favour of any device which will bypass syngamy'. He is alluding, no doubt, to the big element of chance attendant on the water-controlled sexual process in bryophytes. In a word, safety (reliability) in reproduction is secured only at the sacrifice of evolutionary potential. But, one asks, is this completely so? It has long been known, for instance, that chromosomal changes can be induced by colchicine treatment, X-ray bombardment and so on (cf. Heitz,[178] Chopra & Kumar,[82] Hatcher,[168] Kernbach[230]). It is thus reasonable to suppose that from time to time comparable situations may arise in nature. Schuster refers to the possibility of the arctic environment inducing an increased incidence of polyploidy, although Steere[410] found no such situation in a survey he made of mosses from Arctic Alaska. Even after taking account of such possibilities, the general conclusion must be that both monoecism and reliance on vegetative propagation will make on the whole for an unchanging genotype.

Lewis[260] has drawn attention to a number of useful points. He notes the occasional occurrence of 'partial monoecism', with 'intersexual' sex organs (cf. Chapter 8); and he reminds us that autodiploid gametophytes of dioecious mosses will be of the same sex as the haploids whence they came. He also underlines the essentially wasteful nature of dioecism and in this connection recalls that plants of the two sexes may turn up in very unequal numbers. Longton & Greene[265] have brought out the same point in their work with *Pleurozium schreberi*—a common dioecious moss which throughout much of its range is rarely found with sporophytes. Their work illustrates the truth that only prolonged investigation will suffice to provide the explanation of a fact that had long been puzzling. In this instance the decisive factor was the rarity of plants bearing male inflorescences.

Thus we see that the reproductive biology of bryophytes is a subject with many distinct facets. One concerns the distribution of the sex organs; another the effective working of the cytogenetic equipment. In exerting their long-term influence both will operate through spore dispersal; and on the width and scope of this, among other things, many of the facts of geographical distribution will depend. At the same time, both cytogenetic information and the facts of geographical distribution throw light on the slow process of evolution. In the investigation of this process, as one moves on to a bigger time scale, one inevitably calls in the evidence supplied by the palaeobotanist. To some extent all these studies tend to be carried out in separate compartments, for we may never know the cytology of even the best preserved fossils and the plant displaying disjunct distribution has commonly left no hint as to what intervening areas it once occupied. Moreover, different bryologists come to specialise in one branch or another. Yet, in a sense, the strands from all these topics are interwoven in a single complex fabric, the fabric of evolution. What morphologists past and present have managed to make of this we shall see in our final chapter.

I 2

TAXONOMY AND EVOLUTION: CONCLUSION

At the beginning of this book outlines were given of the classification of both liverworts and mosses in order to provide fixed points of reference against which could be seen the considerable wealth of morphological detail that followed in subsequent chapters. Inevitably, evolutionary questions have come up for discussion at various points in the text, not only in the parts that have concentrated on comparative morphology, group by group, but also in the sections dealing with geographical distribution, the fossil record and cytogenetics. It might therefore seem superfluous to add a final chapter which claims for its chief concern the subjects of taxonomy and evolution. We return to these topics, however, for a special reason, namely to examine the extent of the relationship between them. Also, now that diverse aspects of the Bryophyta have been passed in review, it is fitting to look back and attempt some kind of 'summary statement'. Any conclusions which we draw must reflect our views on contemporary classifications and must at the same time take into account the long evolutionary history of the group.

All biological classifications tend to be hierarchical. That is, they are based on a series of categories, each one a smaller, less comprehensive entity than the one that precedes it. The most important in a long series are division (Bryophyta), class (e.g. Musci), order, family, genus and species. Below the species come the infra-specific categories, sub-species, variety, *forma*. Since, in the division Bryophyta, we are dealing with a relatively stereotyped and circumscribed pattern of structure and life cycle, it follows that in all the lower categories of the hierarchical sequence we shall perforce be considering what are little more than minor variations on a common theme. In these circumstances the greatest difficulty always arises in drawing the necessary boundary lines. A glance at the history of bryophyte classification bears this out. It shows us that a century or so ago many boundaries, especially at ordinal, family and generic levels, were quite different from those we know today. The changes, moreover, have gone hand in hand with what might be called a revaluation of the concept of all these three categories. We hinted at this in Chapter 1; we may now illustrate it a little more explicitly.

True, the number of orders of liverworts has not altered greatly since the time of Cavers and Macvicar, but far more families are recognised today than fifty years ago and hardly a year passes without the erection of several new liverwort genera. Most often the new genus houses organisms that have long been known, although sometimes (e.g. in such cases as *Pachyglossa*[183] and *Phyllothallia*) it represents a genuine scientific novelty. The situation in mosses underlines this trend more emphatically. In William Wilson's *Bryologia Britannica* (1855)[469] we find 'Bryaceae' used as an inclusive category for all British mosses except *Sphagnum* and *Andreaea,* although this so-called 'order' was subdivided into thirty-six 'suborders' which contained ninety genera in all. Today our current check-list of British mosses arranges Wilson's Bryaceae in fifteen orders, thirty-nine families and 160 genera. In particular sections of the classification the increase in the number of genera has been much more dramatic. Wilson enumerated ninety-three British species of *Hypnum*. Today only seven remain there, and the rest are distributed in some thirty other genera. To turn still further back, Steere[409] has reminded us that when Hedwig, in 1801, contented himself with a mere thirty-five genera for all the world's mosses known at that time he had his critics, who accused him of separating genera on points of structure that were too minute and too difficult to examine. This was hardly surprising since they would have grown accustomed to the system of Linnaeus, who in the *'Species Plantarum'* of 1753 had listed only eight. When Steere wrote (in 1947) the number had mounted to over 700.

These few remarks will have sufficed to emphasise that all the categories in the taxonomic hierarchy are man-made and that the concepts that govern their making change with the passage of time. This is true also of infra-specific categories, which enjoyed a 'heyday of proliferation' some sixty years ago, when Warnstorf was recognising an excessive number of varieties and *'formae'* in so many species of *Sphagnum,* and others were equally assiduous in putting names to the innumerable variations of form that a keen eye could distinguish within many of the species of *Drepanocladus*. We need to remind ourselves too that the most crucial category of all, the species, is in the last analysis the product of the species-maker. The criteria which are used to fix its limits have always been notoriously difficult to define; and the 'species' of one generation of bryologists will not necessarily be those of the next. Some fifty years ago forty-five species of *Sphagnum* were recognised as British; current check-lists give only thirty. Where Dixon knew (and thought he understood) two species (and five additional varieties) within the '*Plagiothecium denticulatum* complex', today's bryologists have little difficulty in distinguishing the eight British species currently accepted.[151] In a recent paper Crundwell[94] draws our attention to the fact that Dixon, in 1896, was the first to employ the subspecies concept in British moss classification; but he tells us too that of Dixon's very numerous subspecies 24% are now regarded as varieties, 76% as full species. Crundwell sees the subspecies as a potentially useful category which bryologists have been unwise to drop.

Usefulness is indeed a valid criterion in taxonomic matters. In the face of the almost limitless diversity of nature we need an arrangement of living organisms that will make smooth the path of the user. The whole question of what may be called the philosophy of taxonomy has received full treatment elsewhere, and with special reference to groups other than bryophytes (cf. Davis and Heywood[97]); it would be out of place to enter into a detailed discussion here. Suffice it to say that man possesses a natural instinct to classify objects (of whatever kind) and at the same time there is an imperative need for a clear-cut and usable classification (of bryophytes as of other organisms) so that a state of order may be imposed and every kind of user may be served. This, perhaps, is the primary justification for the work of the taxonomist. In whatever scheme emerges he is likely to begin by satisfying his urge to group together those entities which most closely resemble one another. The distinct entities will probably be species and their close similarity will be reflected in a common generic name. This essentially taxonomic activity can operate quite independently of any thoughts on evolution. In the late eighteenth and early nineteenth centuries indeed it did so. Phyletic considerations were superimposed later.

The taxonomist has a second function to discharge. He must give to each recognisably distinct 'kind' of organism a name that will be generally acceptable and as stable as possible. Linnaeus, with his introduction of the binomial system, set the pace in this direction, and the first letter of his name still stands as the authority cited after the latin names of many species, plant and animal, today. Workers with mosses, however, have arbitrarily taken 1801 as the starting point for their nomenclatural activities, and accordingly the name of Hedwig stands as the earliest 'authority' where mosses are concerned. Liverwort names go back to Linnaeus although, as Schuster[384] points out, many of his names had been supplanted by the middle of the nineteenth century. All activity in this branch of taxonomy is governed by the *International Rules of Botanical Nomenclature*, a volume which is periodically revised to accommodate new provisions agreed upon by systematists from all over the world. This work deserves to be better known than it generally is among students. Closer acquaintance with it brings home to us the intricacy of the subject, and at the same time helps us to see how historically conscious the nomenclatural taxonomist has always to be. We also come to understand why it is that stability of names, one of the avowed objectives of nomenclature, is in practice so imperfectly attained.

The history of the naming of bryophytes over the past 250 years can make fascinating reading. Schuster[384] has outlined the events, with ample citation of the leading authors concerned, for liverworts; and as we read his account we quickly appreciate the complexity entailed, a complexity moreover which grows greater as each succeeding generation builds upon what has gone before. He is not himself very encouraging and admits the appalling state of confusion which exists in many sectors at the present time. A principal cause of this confusion lies in the fact that so often, perforce, systematic work involving the description of many apparently new species has had to be undertaken with only an imperfect knowledge of what has gone before. Hence anyone who undertakes a monographic revision finds synonymy on an almost unbelievable scale. For this, or for other reasons connected chiefly with historical events such as priority of usage, he has to alter names; and the goal of stability is at once defeated. The monographer may decide for quite other, purely taxonomic reasons, that name changes are necessary. For example, he may see cogent reasons for subdividing a particular genus into two or three; or a certain species may be found to transgress the limits of the genus in which hitherto it had been placed. In a word, nomenclatural work becomes inseparable from broader taxonomic questions, and in making adjustments to names the specialist soon finds himself revising the limits of broader categories as well.

Thus, there stands before us the goal of permanence and stability. We see in the mind's eye a 'Utopian' state in which all genera stand boldly circumscribed and fully monographed, all regions known floristically—with up-to-date flora or annotated list on which all can rely. At first inspection, such an ideal state might seem relatively easy of attainment. In an earlier chapter we have noted how descriptive work is proceeding apace in many countries (cf. p. 153) and the list given there could be greatly extended. Yet, we know that in a sense the goal is as far off as ever. This arises partly from the nature of nomenclatural and taxonomic practice which, as we have seen, whilst striving for stability results all too often in change. It arises partly from human fallibility and the imperfect knowledge of even the most erudite workers in this field. It arises also, and pre-eminently, from a third cause, namely the fact of evolution.

Because evolution has been and still is a reality we find, all too frequently, not neatly circumscribed and clearly defined entities, but a mutability and intergrading of organisms. In the Bryophyta, this is particularly noticeable in certain genera of Jungermanniales (among liverworts) and Bryidae (among mosses). In the former, *Lophozia* and *Scapania,* in the latter *Bryum* and *Pohlia, Brachythecium* and *Drepanocladus,* illustrate the point, to mention only a few of the many possible examples. In all these instances we are dealing with genera where the boundaries between some of the species are too ill-defined for comfort. In the case of the mosses there is evidence, too, that hybrid sporophytes may be a good deal commoner than has generally been recognised; yet in the liverworts, as Crundwell reminds us, no undisputed hybrid seems ever to have been found in the wild state. Even so, a sharp differentiation between species may be very difficult to achieve. We saw, in Chapter 11, how an understanding of cytogenetics could throw light on particular situations. What we are witnessing therein is the mechanism of short-term evolution.

Evolution in the long-term sense is a history of extinctions, of survivals and of change within these survivals. The details of the change we can never know perfectly but it is likely that, in many instances, what survives today has departed so far from its own direct ancestor of, say, Carboniferous times, that the link between them is no longer recognisable. Elsewhere, to judge from such scanty information as the fossil record has been able to supply, a given form seems to have remained almost stationary, in the evolutionary sense, from that day to this. One thinks of the Carboniferous *Treubites* and living *Treubia* (Fig. 23); of the still earlier *Pallaviciniites* and living *Pallavicinia.* Among bryophytes, it is possible to speak on the one hand of 'living fossils', on the other of 'rapidly evolving groups'. The vastly different rates of evolutionary change implied in such contrasted examples constitute one of the great enigmas of evolution. Often it is easier to trace with some confidence the history of particular organs, or even of gametophyte and sporophyte (considered as separate evolutionary problems) than it is to embark on the much more speculative task of establishing the phylogeny of whole organisms. Much evolutionary discussion has traditionally centred on organisms considered in groups, the groups being the man-made categories of species, genus or family. Clearly, to do this is to introduce fresh assumptions and to complicate the task still further. Every time an evolutionary argument centres in this way on the origin of some genus or family, then it carries with it the assumption that all the members of that genus or family shared, at some time in the remote past, a common origin. This in itself is a big assumption.

If we compare the contemporary outlook on these questions with that prevailing half a century or more ago, we see that a big change has taken place. The bryologists of an earlier generation appear to us to have been naive in their readiness to produce 'phylogenetic trees' and detailed expositions of evolutionary events. Cavers's 'Interrelationships of the Bryophyta'[74] was typical of his period. It was a confident statement with conclusions based upon what seemed a closely reasoned argument. The interpolation of a new generation—the sporophyte—seen at its inception in *Riccia,* and mounting to a climax of near-independence in *Anthoceros,* this was the pillar on which much of the argument rested. Knock away this pillar, as did Wettstein, R.,[461] Evans and others, and postulate instead two similar, radially symmetrical, green generations in the life of the primitive bryophyte, and the whole argument falls to pieces. Once this second viewpoint is adopted, an important part of what Cavers wrote is stripped of all meaning. For he was a staunch supporter of the antithetic theory which had been so eloquently put forward by Bower in *The Origin of a Land Flora.*[47]

In the years between 1918 and 1948 a measure of confusion obtained on this, the central question of the manner of origin of bryophytes and the broad outline of their evolution. There seemed to be no common ground between the two diametrically opposed viewpoints. On balance, the homologous theory gained the ascendancy. In 1948 Fulford[129] turned again to the big question of interrelationships among Hepaticae. She was fair to every point of view but nowhere did she point the direction in which she herself believed the truth to lie. She provided a most timely and useful review of the literature but pronounced no judgment. Unfortunately, no comparable review exists for the Musci. Less diverse than liverworts, they have all along attracted less interest in this evolutionary context, and in some ways present even more of an enigma.

Within recent years there have been sporadic revivals of interest in the interpretative morphology of bryophytes. Conspicuous in this field has been Mehra,[286] who has put forward a carefully reasoned argument to account for the origin of the complex, chambered thallus of the 'higher' Marchantiales from such Metzgeriales as *Petalophyllum* (cf. Fig. 1, p. 25). Both the idea and the argument are ingenious, but to many bryologists the theory will not be wholly convincing. In quite a different direction, Christensen[83] and Haskell[166] were among those who startled some morphologists by their revolutionary attempt to trace the descent of bryophytes from the simplest known vascular plants—the Psilophytales. This viewpoint has tended to gain ground in the last few years (cf. Schuster[384]) and it is well to make sure we understand just what is implied.

There springs to mind the notion of bryophytes having 'evolved downwards' from well-known living pteridophytes, but this is not what is meant at all. The postulated ancestors, according to this third theory, are in fact the unknown antecedents of the most primitive known fossil pteridophytes, and must therefore have been plants which were themselves barely on the threshold of pteridophyte status. Also, the simplest members of the extremely ancient but well preserved Psilophytales were themselves not quite so far removed from *Anthoceros* as might at first be imagined. True, they were vascular plants, but the genera *Rhynia* and *Horneophyton* lacked both roots and leaves. There was a creeping axis which bore rhizoids and sent up at intervals erect, cylindrical, dichotomising branches which bore terminal ovoid–cylindrical sporangia (Fig. 25B). A columella was present in *Horneophyton* but lacking in *Rhynia*. Thus, it is perhaps not altogether far-fetched to seek the origin of bryophytes in forms midway between these and the sporophytes we know in modern Anthocerotales. It is tantalising of course that we know nothing of the gametophytes of these Devonian fossils; nor has the fossil record provided us with any clues as to the early history of *Anthoceros* itself.

Because the early fossils of bryophytes are so few, the time-span so long, and because organs or generations (sporophyte, gametophyte) have apparently undergone change at differential rates, many would say (cf. Smith & Newton[402]) that conjecture on the early evolution of the group amounts to little more than futile speculation. Yet there are others who do not hold this view. Anderson[9] has seen the elucidation of evolution as a principal aim of systematics. Fulford[134] has herself returned in recent years to a new evaluation of data which might have some bearing on our overall interpretation of relationships within the Hepaticae. She reminds us that Kashyap[228] long ago postulated a pteridophyte ancestry for bryophytes and she seems ready to acquiesce in Proskauer's more specific assertion of a direct link between *Horneophyton* and *Anthoceros*. Fulford[134] believes, however, that as yet too few genera have undergone full morphological investigation and that only after many gaps in our knowledge have been filled can we hope for 'an objective interpretation of the position of the hepatics in relation to other groups of plants'.

One is left wondering to what extent Fulford is prepared to consider the possibility of a polyphyletic origin for bryophytes. Schuster[334] leaves us in no doubt as to where he stands, for he makes clear reference to 'possible polyphyletic evolution from erect gametophytes'. He agrees with Fulford that there are as yet far too many liverworts imperfectly investigated, but notwithstanding these gaps

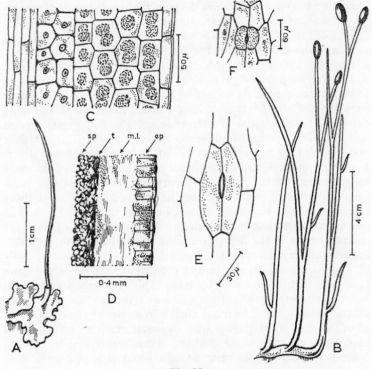

Fig. 25

Certain attributes of *Anthoceros* and *Rhynia* compared. A. Part of plant of *Anthoceros husnotii* with 'horn-like' capsule emerging from collar-like involucre. B. Part of reconstruction of fertile plant of *Rhynia gwynne-vaughani* (adapted from Boureau et al.). Note absence of roots and leaves; dichotomously branched stems and terminal sporangia. C. Part of longitudinal section of capsule of *Anthoceros* sp., showing from left to right, the four layers of narrow columella cells, sporogenous tissue (as yet only two layers thick), four layers of chlorophyllose wall cells, and epidermis. D. Part of longitudinal section of the sporangium of *Rhynia gwynne-vaughani* (after Kidston and Lang). ep. epidermis; m.l. disorganised middle layers of wall; t. tapetum; sp. spores. E. and F. Surface views of stomata, E. from capsule of *Anthoceros* sp., F. from stem of *Rhynia gwynne-vaughani* (after Kidston and Lang).

in our knowledge, he takes a firm stand on several fundamental points. One is the derivation of all dorsiventral gametophytes from more of less 'erect, radially symmetrical isophyllous prototypes' (an idea which he traces back to Wettstein, R.[461]). Another is his insistence that the dependent or 'parasitical' mode of life of the sporophyte has brought with it profound change and especially a long-term reduction in complexity in that generation. He also sees numerous 'reduction lines' as operative in the gametophyte generation, notably (as we saw in Chapter 3) in the Jungermanniales.

Schuster is able to quote Schwarz,[389] Heintze[177] and Jennings[217] as others who have favoured the idea of bryophyte origin by way of a simplification of the morphology exhibited by the Psilophytales; and he throws in the weight of his own vast experience and authority to support this view. At the outset such a 'simplification' will have been far reaching, for it must have entailed total loss of branching capacity as well as loss of independence. The downgrade process will then have continued *within* the Hepaticae to reach its ultimate expression in the sporophyte of *Riccia*. Some such sequence strikes him as far more inherently probable than the alternative, favoured by supporters of the antithetic theory, viz. the intercalation of a 'new' sporophyte generation (along the lines of *Riccia*) at a relatively late stage in time. He readily admits, however, that whatever scheme one devises in an attempt to express a possible course of events, the *ultimate* origin of the Bryophyta must always be seen to lie within the green algae (Chlorophyta). It is thus only a question as to exactly at what point the Bryophyta and the Psilophytales diverged from their hypothetical common ancestor. This is a broad conclusion but nevertheless an important one; and the student will notice that a modern general textbook (Bell & Woodcock[22]) wisely reminds us of the same truth, namely that—however much we may argue as to detailed pathways—the evidence is irrefutable that in the last analysis the bryophytes came from a green algal ancestry. With green algae they share (loc. cit.) photosynthetic pigments, cell wall components, food reserves and flagellar characters. Moreover, if this view is accepted, then we see perhaps in the filamentous protonema of modern mosses a 'recapitulation' of their ancestral adult form.

Clearly, to take the discussion so far, and no further, is to leave a great many important questions unanswered. Can we, for instance, postulate in general an evolution from filamentous, through thalloid to leafy gametophyte form? Such a morphological sequence would be compatible with our knowledge of events within various classes of algae, where a thalloid plant body appears as an advance on a filamentous one and where no leafy forms are known. Yet current views on evolution within the Hepaticae would not seem to permit any such broad generalisation, for leafy gametophytes therein are widely conceded to have given rise secondarily to thalloid forms. Or again, even if we agree that *Anthoceros* sporophytes come closer than those of any other living bryophytes to the sporophytes of hypothetical 'pre-Psilophytalean' ancestors, what knowledge have we of the immense gap that must separate the latter from the diploid generation of a branched filamentous alga exhibiting an isomorphic diplobiont life-cycle? The answer must be none at all. Then, turning again to the gametophyte generation, although we have no knowledge of the gametophytes of early fossil Pteridophyta, such as the Psilophytales, we can see some suggestive parallels between the gametophytes of some *living* Lycopsida and that of the liverwort *Anthoceros*; and there are some striking resemblances, too, between the archegonia in these two cases, for in both they are reduced in size, immersed in (or confluent with) surrounding tissues and borne on the upper surface of a thalloid gametophyte. Do such observations strengthen the case (already strong on grounds of chloroplast structure, etc.) for regarding the gametophyte of *Anthoceros* as also being close to the primitive ancestral form, notwithstanding the fact that it is neither leafy nor radially symmetrical? All such questions are more easily asked than answered, and different morphologists will always tend to hold divergent views upon them. Reluctantly, we are forced to agree with Schuster[384] when he writes 'A solution to the problem of the origin of the Hepaticae (and of the mosses) and the initial steps in their evolution appears to be nearly as remote as it was some fifty years ago'. If origin and initial steps remain hidden from us we can scarcely hope to work out effectively the interrelationships within the group.

The situation in the Musci is perhaps even more unyielding than that in the Hepaticae. We are almost completely without clues regarding the links between the more strongly contrasted forms. Genera like *Sphagnum* and *Andreaea* stand very much in isolation. Sixty years ago Cavers[74] saw both originating on a bifurcating evolutionary line which sprang from an unknown common ancestor combining some features of the Anthocerotae with other features of the Jungermanniales. His evidence came largely from sex organ structure and sporophyte ontogeny. Such modern views as have been expressed are very different (see p. 66). We saw in Chapter 11 how little the fossil record could help. Also, the fact that sporophyte and gametophyte seem so often to have followed completely independent evolutionary paths greatly complicates the task of 'placing' whole organisms. We saw this repeatedly in the early chapters of this book; and it applies to both mosses and liverworts. Thus, *Sphaerocarpos* and *Buxbaumia* had apparently simple gametophytes and relatively complex sporophytes; *Riccia* and *Polytrichum* showed the reverse combination. Yet the apparent simplicity of the sporophyte of *Riccia* and the gametophyte of *Buxbaumia* raises another issue. For they are very widely regarded nowadays as illustrations not of primitive simplicity, but of extreme reduction associated with the abandonment of an autotrophic mode of life. This immediately makes one ask another question. If all complex forms are seen as highly evolved and most simple forms are regarded as reduced, then where are the *relatively simple and unspecialised* examples which genuinely mark the stages on the *upward* path of increasing complexity and increasingly perfect evolutionary adaptation? So far as structures like the complex chambered thallus of the 'higher' Marchantiales and the 'perfect' double peristome of moss capsules are concerned, we are extraordinarily short of such examples. This is a point that seems sometimes to have been overlooked.

Thus, to summarise, there has been, over a time-span of more than 300 million years, an evolution of bryophytes, but we find ourselves as yet unable to trace the steps by which it has taken place. Even the most plausible theories—like the antithetic theory of sixty years ago—stand opposed by counter-theories; and really good evidence is in short supply. It is easier to speak of 'levels of organisation' in respect of particular organs than it is to do the same for whole organisms. Often these latter display contradictory features, especially in the two generations, haploid and diploid, which compose the whole. The older morphologists constructed phylogenetic trees, on which they placed whole orders, sometimes stating boldly that one was derivative from another, at other times making a generous allowance for missing links (cf. Leitgeb,[258] Campbell,[64] Cavers[74]). Few bryologists attempt to do the same today (cf. however Gams[137]). Instead, the modern observer admits that he is largely baffled. He sees a hypothetical ancestral form (or forms) somewhere between Chlorophyta and the most primitive known extinct Psilopsida but he knows it to be as yet 'a missing link' unsupported by concrete fossil evidence. He sees certain valuable and informative fossils (as we noted in Chapter 11) which give him direct evidence concerning the antiquity of particular groups. But about their interrelationships he knows next to nothing and so is inclined not to make dogmatic statements. He tends to favour some variant of the 'homologous theory' on the whole, according to which he envisages a hypothetical early bryophyte with the two alternating generations more alike than they are in any known forms, living or fossil; but again, he has never met this plant and so is not unduly worried if some other observer continues (cf. Meeuse[285]) to uphold the antithetic view. When the early pteridophytes are brought more actively into the argument this does not seem unreasonable to him either. He acquiesces in a situation where all are free to express an opinion, but where the morphological evidence is sometimes so equivocal as to be of doubtful value. Much of the heat has gone out of the old controversy.

In view of this obscure evolutionary picture, what is the meeting point, the common ground, between the two—evolutionary studies on the one hand, taxonomic practice on the other? In short, how far can taxonomy reflect phylogeny? Quite clearly, it can do so at most only very imperfectly. In the first place, a systematic list is by its very nature linear, uni-dimensional; whilst evolutionary events are three-dimensional, and even if these events were known they could never be expressed fully in a systematic list, however skilfully that was arranged in genera, families, orders and so forth. Secondly, the 'missing links' are far too many; they confront us on every side. For the most part we are looking at a tree of which we can see clearly only the extremities of some of the branches. These and other considerations incline us to urge that taxonomic practice must proceed on its clearly defined task quite independently of any phyletic implications. We begin to understand the point of view of those who would divorce the two completely.

Yet we have no sooner confronted this situation, than we see how untenable such a view must be. Taxonomy, as we have seen, supplies the need for reliable names and an orderly arrangement. The first we see reflected in such important contemporary works as the *Index Muscorum*[431] and *Index Hepaticarum*;[38] the second is reflected in every carefully prepared check-list or flora. Taxonomic activity, however, also springs from our urge to group similar objects together, to place like with like. In so doing, given a belief in evolution, we are inevitably placing together those organisms that we believe to be more or less closely related genetically; equally, and by the same token, we separate widely those we believe to share a common ancestor only very far back in time. Faced with the diversity of bryophytes, we make our taxonomic judgments on such evidence as is available. We cannot escape the fact that they are also in some degree phyletic judgments. Even though the sequence of the main events in bryophyte evolution may continue to elude us, every time we place certain species, genera or families together in a taxonomic scheme we are, by implication, declaring our faith in a particular interpretation of some tiny corner of the vast evolutionary picture. To deny this is to deprive our bryophyte taxonomy of a great part of its meaning.

It is well at this point to look a little more closely at the kinds of evidence on which bryologists in practice base their taxonomic judgments. A glance back at earlier chapters will suffice to show that,

G

in both mosses and liverworts, some characters have traditionally been seen to be more important than others. Thus, among leafy liverworts (Jungermanniales) the insertion of leaves on the stem (whether succubous or incubous), the structure of the perianth and the composition of the capsule wall—all these have been regarded as characters of importance. The same may be said of the overall 'pattern' of leaf cell structure and the architecture of the peristome in mosses. It is as well to be clear about the reasoning behind this kind of judgment. In effect, the taxonomists of the past, in making a decision of this kind, have affirmed their belief that such characters as these are inherently less likely to undergo change through natural selection pressures than are characters of another class, such as leaf size, habit, colour and so on. The difference of course is that between the fundamental and the superficial. There are those today who question whether one character is any more important than another and who believe that comparisons should be based upon the numerical sum of the attributes shown by two organisms, no more. Yet to concede this is to deny the validity of taxonomic judgment; the judgment which has given us our hierarchical system of classification. A major group (such as the Andreaeidae, among mosses) is set apart from another major group (the Bryidae), not because an exceptionally long list of differences can be found between them, but because certain characters such as the development and mode of dehiscence of the capsule and the structure of the antheridia proclaim the fundamental isolation of *Andreaea*. The judgment of the taxonomist is used to distinguish between the fundamental and the superficial; and this has always been so, in the systematics of every class of organisms. Sometimes, indeed, the 'underlying' characters which can help us to reach a decision are concealed from view. This applies to the facts of cytology and of phytochemistry, both of which call for special types of investigation to reveal them.

We have seen how characters such as these (cytological, phyto-chemical) could be employed in particular cases in the role of arbiters. Smith and Newton[402] made a cytologist's judgment regarding the taxonomic treatment of *Tetraphis*; Lewis, D.,[259] used phyto-chemical evidence to assess the validity of the new placing of *Plagiochila carringtonii*. Few would deny that in both these instances we are dealing with characters that are fundamental. Indeed, it is common experience to find that plants which are genuinely unrelated are distinguished from one another by a considerable range of characters, all of a kind to which some importance can be attached. Among bryophytes, cases where a strong superficial resemblance between quite unrelated plants has been engendered through a kind of 'parallel development' are rare. The common oceanic moss

Hookeria lucens bears a certain resemblance in its leafy shoots to some leafy liverworts and in quite a different way the same may be said of *Hypopterygium* (with its leaves strictly in three ranks), but these superficial resemblances do not extend to the rhizoid system, still less to the sporophytes, so that no possible confusion can arise as to the true position of the plants. These and other examples, however, will serve to emphasise the wisdom of always taking into account the sum of the available characters. While admitting a distinction between the superficial and the deep, we need to make use of all the evidence we have. When relatively new techniques, such as those of phytochemistry, provide a fresh body of information, we must welcome it and assess it in relation to what has gone before.

Whenever we speak of a particular classification as being sound we imply, among other things, that it places together those entities which are *genetically* not far apart. This conclusion is inescapable. Thus, to cite actual examples, every classification of mosses places *Bryum* and *Pohlia* near one another; all liverwort arrangements show *Gymnomitrion* and *Marsupella* side by side. This placing affirms a belief in intergeneric relationship in each instance. What is impossible is to assess the degree of genetic relationship between taxa placed far apart in our classification. *Corsinia* and *Anthoceros* are both thalloid liverworts; *Polytrichum* and *Funaria* are both genera of acrocarpous mosses; but there, in each instance, the resemblance ends and we have absolutely no means of knowing the extent of true phyletic relationship between the members of either pair. We are thus compelled to admit that there are strict limitations to the extent to which taxonomy can reflect phylogeny. Yet a certain phyletic implication is written into every classification of living organisms; and when an author erects a new order for some new discovery (as happened in the case of *Takakia lepidozioides*) that step is a measure of his conviction that in the *evolutionary* sense it stands far apart from everything hitherto known.

At this point it will be useful to turn back to actual examples in a little more detail, in an attempt to answer the question which many students will be asking. Amid such contradictory views and theories regarding the broad course of bryophyte evolution, can we state with any conviction what are the most primitive among living forms of liverwort and moss? In the Hepaticae the following may in different ways be regarded as exhibiting some outstandingly primitive features: *Takakia*; *Haplomitrium* (Fig. 26); *Anthoceros* and its immediate allies; a range of radially organised genera of the Jungermanniales. Cronquist's[91] conclusion that *Anthoceros* displays primitive features in its sporophyte, advanced ones in its gametophyte, strikes one as arbitrary. It springs from a too rigid adherence to the 'homologous' theory and is rendered unnecessary if one inclines to see some form of descent from early pteridophyte ancestors. It is, incidentally, the opposite view to that held by morphologists of an earlier generation. Clearly, however, these examples are primitive in very different ways. Some Metzgeriales, as we have seen from the fossil record, are undisputably ancient; for this reason, and on grounds of comparatively simple thalloid structure, they too may be accounted primitive, although not all would support this verdict (cf. Schuster[384]). *Monoclea* may well be an ancient form but there is no fossil evidence either to support or to refute this view. Among 'advanced' liverworts we may cite families such as Lejeuneaceae and Frullaniaceae in the Jungermanniales; possibly also the most complex Marchantiales, with their highly specialised thalli and carpocephala. At best, however, these are tentative conclusions, and in making them we are not helped to see more clearly the actual course of liverwort evolution.

In the Musci, again in different ways, various isolated and aberrant genera may perhaps be looked upon as primitive survivals. At least each has been so regarded, at some period in the history of the subject, by some morphologists. The principal examples that spring to mind are: *Sphagnum, Andreaea, Polytrichum* and its allies, *Tetraphis* (and the related *Tetrodontium*), *Archidium, Buxbaumia*. All have received attention at various points in this text and the student will already be aware of certain considerations which prevent us from reaching any firm conclusions. These are (1) the fact that the first four of these combine some primitive features with others that are clearly very specialised; (2) the other two show strong indications of being reduced forms; (3) the almost total lack of relevant fossils. Some (but not all) earlier observers certainly cast *Archidium* and *Buxbaumia* in the role of primitive mosses. Consulting Campbell,[64] one appreciates afresh how far back these disputes go; also how heavily the early workers leaned on developmental evidence. We have already noted how Cavers[74] used *Sphagnum* as a 'bridge' from liverworts to mosses. It is interesting to note the importance which this very thorough observer attached to the link between a dome-shaped archesporium (as seen in *Sphagnum* and *Andreaea*) and a relatively undeveloped state of the operculum and peristome regions of the capsule. Once the columella penetrated through to the distal extremity it provided (according to him) the means whereby refined differentiation of annulus and peristome could proceed. Thus *Andreaea,* with no such penetration, was another early form; and it was a short step from the four-valved capsule of *Andreaea* to the four 'solid' peristome teeth of *Tetraphis. Polytrichum*

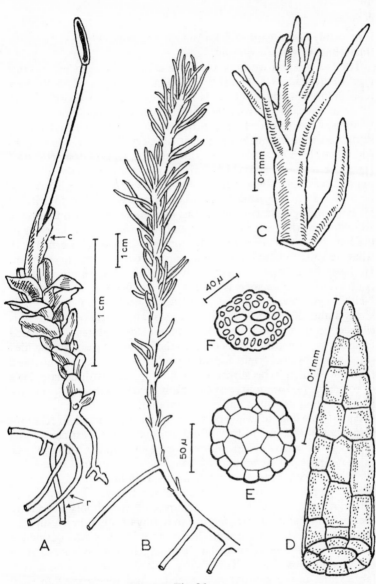

Fig. 26

Haplomitrium and *Takakia*. A. *Haplomitrium mnioides*, fertile plant
(after Schuster). c. calyptra; r. root-like branches of axis. B–E. *Takakia
lepidozioides*. B. Axis, 'phyllids' and root-like branches (after Schuster).
c. Tip of shoot and D. single phyllid (of simple but solid construction)
from plants provided by W. B. Schofield and grown at Reading. E. Transverse section of phyllid of more complex construction (after Schuster).
F. *Takakia ceratophylla*, T.S. phyllid (after Grolle).

is probably best regarded as far out on a 'side-track', away from the 'main line' of moss evolution.

The few modern workers who have expressed clear views on this type of question (cf. Gams,[137] Khanna,[232] Chopra, R. S.[81]) have tended to stress different kinds of evidence and to arrive at very different conclusions from those of the early morphologists. Gams minimises the so-called unique features of *Sphagnum* and arrives at the remarkable conclusion that it is a derivative of the big 'Arthrodont' order Dicranales. He draws up the type of 'phyletic tree' (rarely seen nowadays) which makes a direct evolutionary sequence out of a series of living genera. Thus he links *Tetraphis* with *Schistostega* by way of *Tetrodontium*. Although one is inclined to criticise his bold and speculative conclusions one values his return to a topic so seldom worked over nowadays and the broad review he gives of the long history of the subject. Khanna is especially concerned with what he calls 'differential evolutionary activity' in the bryophytes. He lays himself open to numerous criticisms, but although highly controversial, his paper is useful for its breadth of viewpoint. Neither cytogenetic, nor fossil evidence is neglected. Chopra[81] also stresses the differences between mosses and liverworts and it is instructive to see how very different are the features emphasised from those on which Cavers chiefly relied. Chopra discards the idea of a monophyletic origin for all bryophytes. If we accept his view then we shall not be surprised to find several evolutionary lines of roughly equal antiquity; or to see alive today numerous 'primitive survivals' that are quite unlike one another.

Among the lines of evidence which tell us not so much of relationships, but of the past history of organs, one of the most convincing is that provided by vestigial structures. For example, the existence in the capsule of *Sphagnum* of non-functional stomata proclaims for that capsule an earlier existence as a photosynthetic organ of greater importance than the one we see today. The apparently functionless underleaves near the apex of the shoots in certain species of the family *Lophoziaceae* can hardly be other than a survival from a past in which these plants had a ventral rank of fully developed leaves. This kind of evidence has been largely used by advocates of the homologous theory. This theory calls for reduction on a very large scale, in the sporophyte especially, and *Riccia* is often cited as the end product of the process. In the absence of any known sporophyte in bryophytes with even a trace of appendages, however, it is not easy to be convinced by an argument which postulates two originally similar generations. This is perhaps one reason why morphologists have looked in the direction of pteridophytes for the ancestors of the Bryophyta.

If it is true that many problems of interpretative morphology must remain at present unsolved, it is equally true that recent decades have seen a resurgence of interest in bryophytes from many points of view. They seem unlikely ever to achieve great economic importance and hence attract the limelight which that entails. Yet among recent studies are some which have pointed to the value of bryophytes as indicators of radioactive fall-out,[387] and we have seen earlier (Chapter 10) how in different cases their presence or absence can provide valuable evidence concerning such diverse topics as atmospheric pollution[141] and copper concentrations in the soil.[394] There has also begun a not entirely unsuccessful search for antibiotics in the group (McCleary & Walkington[273]); and it is difficult to forecast what substances of future significance might in time be revealed by the phytochemists. Detailed knowledge of the lives (and especially the physiology) of individual species is still in its infancy. Bryophytes appear to be susceptible to attack by a comparatively small range of predators; but many thalloid liverworts are in varying degree mycorrhizal. There is vast scope for further experimental work in these fields.

One needs to know with what species one is working; so the taxonomist will always be needed. He cannot achieve finality but he can provide what currently seems to be the closest approximation to truth. One of the greatest needs of the present day is the revision of large genera which include numerous tropical representatives that have possibly never been submitted to critical reassessment since the date of their original description. The task is formidable and many of those attempting it have confined themselves of necessity to particular regions. The total output of such studies, over the past twenty-five years, is impressive. We can cite only a few examples. Thus, there is the work of Fulford[128] on *Bazzania* in Central and South America, followed by a much more wide-ranging study of Jungermanniales as a whole.[133] Schuster has been active on a very broad front; and this has included monographic work in many groups of Jungermanniales, also in the Metzgeriales (especially of southern latitudes) and the Calobryales.[385] For some thirty years Castle[69] has been pursuing his study of *Radula*; rather similarly Clark earlier pursued her long-term study of *Frullania*. Grolle[155] has monographed genera and families in many parts of the world, whilst Jones[224] has made over twenty contributions on the hepatics of tropical Africa, unravelling many 'taxonomic knots' as he did so. Then there have been Bischler, Bonner & Miller[33] at work on various Lejeuneaceae, Hatcher[167] on *Isotachis,* and, more recently, Kitagawa[234] on the Lophoziaceae of Japan.

Among mosses, genuine monographic revisions of particular

systematic groups have perhaps been fewer. From Japan have come the work of Noguchi[307] on Leucodontineae and Neckerineae (of the big family Neckeraceae), of Ando[11-13] on *Hypnum, Homomallium* and other genera, and of Ochi[311] on Bryaceae. More recently Takaki[418] has monographed the Japanese components of several genera of Dicranales, and Osada[314] has done the same for Polytrichales. Shin[395] has undertaken the formidable genus *Fissidens,* and so on. One sees here, surely, constituent materials for a future moss flora of Japan. Meanwhile, Reese[355] (*Calymperes*), Lawton[249] (various Hypnobryales) and Welch[455-6] (Fontinalaceae, Hookeriaceae) are among those who have been active on the American continent. The handling of moss collections from remarkably diverse regions was the special interest of Bartram (some forty papers between 1945 and 1960), although bryologists from all over the world have of course played their parts in this direction The challenge of such work is formidable and there is a constant need for close co-ordination with what has gone before. The position has been well stated by Verdoorn[433] in an article of some years ago, '*The future of exotic cryptogamic botany*'. One big question is which should come first, the monographic revisions or the definitive flora. Without the former, the latter can scarcely be very complete or reliable.

The moss flora of Western and Central Europe has been reasonably well-known for a long time (cf. Husnot,[199] Limpricht,[261] Roth,[369] Monkemeyer;[297] the liverworts have, in recent years, been the subject of Müller's[300] comprehensive study, whilst Augier's[19] splendid volume deals with both. Species concepts change, however, and fuller details of distribution are worked out. This type of investigation, indeed, is still very much alive—witness the 'crop' of important contributions from many countries within the last ten years. The year 1961 saw the publication of the first volume of the moss flora of Arctic Russia (Abramova et al.[1]). In Eastern Europe Bulgaria (Kúc, Vajda & Pocs), Czechoslovakia (Duda & Vana[114]), Hungary (Boros[44]), Poland (Kúc,[242] Szafran[417]) and Roumania (Papp) have all received attention. Important bryophyte floras have appeared for Scandinavia (Arnell[16], Nyholm[309]) and for Belgium (Demaret & Castagne[100]). Indeed no country has lacked investigators and the most urgent modern need might be said to be co-ordination.

Many British bryologists could with advantage be better informed regarding the Continental distributions of the plants they study. Within the British Isles there has been a long tradition of interest in geographical distribution, but in the last decade emphasis has shifted somewhat from the vice-county to the 10km square. Between 1962 and 1970 distribution maps of sixty-three mosses and sixty-one liverworts have been published on this basis in the Transactions of the British Bryological Society; and there have been 'county floras' too numerous to mention individually, some twelve in all since 1945 (cf.[220,323,339,367,416]). In the first edition it was remarked that over a dozen mosses and several hepatics had been found new to Britain between 1945 and 1962. In the past eight years a further sixteen mosses and eight liverworts have been added. Some of these result from a closer scrutiny of critical groups;[95] others are in the nature of alien adventives; still others represent important range extensions (cf.[330,442]). There might have been some rash enough to suggest, thirty years ago, that the distribution of bryophytes in Britain was fully 'worked out'. The challenging discoveries of the past twenty-five years have given the lie to any such notion.

One final word to the beginner. In this book I have not hesitated to give extensive references to the literature, in the hope that the student will be led thereby to explore in depth, utilising original sources. Yet the beginner must not expect any easy or rapid conquest of a difficult subject and he will surely be wise to proceed by way of general textbooks such as those of Smith and Parihar, through review articles, to the original papers; just as he may wish to begin with a simplified flora[448] and graduate to the handbooks of Dixon[107] and Macvicar.[277] At first he may be inclined to despair of ever learning to recognise even quite common species. For there is a double difficulty; the 'form circle' of individual species is often surprisingly wide whilst at the same time the boundary between two species can sometimes be fine indeed. It is well to remember that the exceptional eye, combined with taxonomic flair, is given only to the few. Far more than this can be bryologists; and there is much for them to explore. Richards[362] has looked back, down the 'vista' of the last half-century, and has traced many paths of progress. The young bryologist is privileged to look forward, down the long vista that lies ahead. Without doubt, new and exciting discoveries await him.

BIBLIOGRAPHICAL REFERENCES

1. Abramova, A. L., Savicz-Ljubitzkaya, L. I., & Smirnova, Z. N. (1961). *Handbook for the determination of the mosses of Arctic Russia* (in Russian), Moscow, 748 pp.
2. Allen, C. E. (1935). Bot. Rev. **1**, 269
3. Allen, C. E. (1945). Bot. Rev. **11**, 260
4. Allorge, V. (1955). Rev. bryol. lichen. **24**, 248
5. Allsopp, A., & Mitra, G. C. (1958). Ann. Bot. N.S. **22**, 95
6. Amann, J. (1928). *Bryogéographie de la Suisse,* Zürich
7. Andersen, E. M. (1929). Bot. Gaz. **88**, 150
8. Anderson, L. E. (1962). In Altman, P. L., & Dittmer, D. (Eds.), *Growth, including Reproduction and Morphological Development,* pp. 45–57, Washington, 608 pp.
9. Anderson, L. E. (1964). Phytomorphology **14**, 27
10. Anderson, L. E., & Bryan, V. S. (1956). Rev. bryol. lichen. **25**, 254
11. Ando, H. (1958). J. Sci, Hiroshima Univ. Ser. B, Div. 2, **8**, 167
12. Ando, H. (1965). Hikobia **4**, 28
13. Ando, H. (1966). Bot. Mag. Tokyo **79**, 759
14. Andrews, A. L. (1960). Bryologist **63**, 179
15. Andrews, H. N. (1960). Palaeobotanist **7**, 85
16. Arnell, S. (1956). *Illustrated moss flora of Fennoscandia,* 1: Hepaticae, Lund
17. Arnold, C. A. (1932). Pap. Mich. Acad. Sci. Arts & Lett. **15**, 51
18. Arnold, C. A. (1947). *An Introduction to Palaeobotany,* New York
19. Augier, J. (1966). *Flore des Bryophytes,* Paris, 702 pp.
20. Barkman, J. J. (1958). *Phytosociology and ecology of cryptogamic epiphytes,* Assen, 628 pp.
21. Bauer, L. (1963). J. Linn, Soc. (Bot.) **58**, 343
22. Bell, P. R., & Woodcock, C. L. F. (1968). *The Diversity of Green Plants,* London, 374 pp.
23. Bennet, R. A. (1965). Trans. Proc. bot. Soc. Edinb. **40**, 121
24. Benson-Evans, K. (1964). Bryologist **67**, 431
25. Benson-Evans, K., & Brough, M. C. (1966). Trans. Cardiff Nat. Soc. **92**, 4
26. Berrie, G. K. (1958). Trans. Brit. bryol. Soc. **3**, 427

27. Berrie, G. K. (1960). Trans. Brit. bryol. Soc. 3, 688
28. Bhandari, N. N., & Lal, M. (1968). Bryologist 71, 122
29. Biebl, R. (1967). Flora B. 157, 25
30. Birse, E. M. (1957). J. Ecol. 45, 721
31. Birse, E. M. (1958). J. Ecol. 46, 9
32. Bischler, H. (1961). Rev. bryol. lichen. 30, 232
33. Bischler, H., Bonner, C. E. B., & Miller, H. A. (1963). Nova Hedwigia 5, 359
34. Bizot, M. (1965). Bull. Soc. linn. Lyon 34a, 306
35. Black, C. (1913). Ann. Bot. 27, 511
36. Bliss, L. C., & Linn, R. M. (1955). Bryologist 58, 120
37. Boatman, D. J. (1961). J. Ecol. 49, 507
38. Bonner, C. E. B. (1962–). Index Hepaticarum, Weinheim
39. Bopp, M. (1954). Z. Bot. 42, 331
40. Bopp, M. (1959). Rev. bryol. lichen. 28, 319
41. Bopp, M. (1961). Biol. Rev. 36, 237
42. Bopp, M. (1968). Ann. Rev. Pl. Phys. 19, 361
43. Bopp, M., & Stehle, E. (1957). Z. Bot. 45, 161
44. Boros, A. (1968). Bryogeographie und Bryoflora Ungarns, Budapest, 466 pp.
45. Bowen, E. J. (1931). Ann. Bot. 45, 175
46. Bowen, E. J. (1933). Ann. Bot. 47, 401 et seq.
47. Bower, F. O. (1908). The Origin of a Land Flora, London, 727 pp.
48. Briggs, D. (1965). J. Ecol. 53, 69
49. Brodie, H. J. (1951). Can. J. Bot. 19, 224
50. Brotherus, V. F. (1924). Musci. Spezieller Teil. In Engler and Prantl, Die natürlichen Pflanzenfamilien, Bd. 10 and 11, Leipzig
51. Bryan, G. S. (1927). Bot. Gaz. 84, 89
52. Bryan, V. S. (1955). Bryologist 58, 16
53. Bryan, V. S. (1956). Bryologist 59, 118
54. Bryan, V. S. (1956). Amer. J. Bot. 43, 460
55. Buch, H. (1911). Über die Brutorgane der Lebermoose, Helsingfors, 69 pp.
56. Buch, H. (1930). Annls. bryol. 3, 25
57. Buch, H. (1947). Soc. Sci. Fenn, Comm. Biol. 9, 1
58. Buch, H., Evans, A. W., & Verdoorn, H. (1938). 'A preliminary check-list of the Hepaticae of Europe and America (North of Mexico)', reprinted from Annls. Bryol. 10, (1937), Leiden, 8pp.
59. Bunce, R. G. H. (1967). Bot. Notiser 120, 334
60. Burgeff, H. (1943). Genetische Studien an Marchantia, Jena
61. Burges, A. (1951). J. Ecol. 39, 271
62. Burr, F. A. (1970). Am. J. Bot. 57, 97
63. Campbell, D. H. (1898). Bot. Gaz. 25, 272
64. Campbell, D. H. (1918). The structure and development of mosses and ferns, 3rd ed., London, 708 pp.
65. Campbell, D. H. (1924). Ann. Bot. 38, 473
66. Campbell, D. H. (1936). Bot. Rev. 2, 53
67. Carothers, Z. B., & Kreitner, G. L. (1968). J. Cell Biol. 36, 603
68. Carr, D. J. (1956). Australian J. Bot. 4, 175
69. Castle, H. (1968). Rev. bryol. lichen, 36, 5
70. Cavers, F. (1903). Ann. Bot. 17, 270
71. Cavers, F. (1903). New Phytol. 2, 121
72. Cavers, F. (1904). Rev. bryol. 31, 69
73. Cavers, F. (1904). Ann. Bot. 18, 87
74. Cavers, F. (1911). Interrelationships of the Bryophyta. New Phytol., repr. no.4, Cambridge, 203 pp.

75. Chalaud, G. (1931). Annls. bryol. **4**, 49
76. Chalaud, G. (1932). Germination des spores et phase protonémique. In Verdoorn, F., *Manual of Bryology*, 89
77. Chalaud, G., (1937). Rev. gén. Bot. **49**, 111
78. Chen, P. C., & Wu, P. C. (1964). Acta phytotax. sin. **9**, 213
79. Chopra, R. N., & Gupta, U. (1967). Bryologist **70**, 102
80. Chopra, R. N., & Rashid, A. (1967). Bryologist **70**, 206
81. Chopra, R. S. (1968). Phytomorphology **17**, 70
82. Chopra, R. S., & Kumar, S. S. (1961). Bryologist **64**, 29
83. Christensen, T. (1954). Bot. Tidsskr. **51**, 53
84. Clapp, G. L. (1912). Bot. Gaz. **54**, 177
85. Clausen, E. (1952). Hepatics and Humidity. In Dansk bot. Arkiv. **15**, no.1, 80 pp.
86. Clements, F. E. (1916). *Plant Succession,* Washington, 512 pp.
87. Clymo, R. S. (1963). Ann. Bot. N.S. **27**, 309
88. Clymo, R. S. (1970). J. Ecol. **58**, 13
89. Coker, W. C. (1903). Bot. Gaz. **36**, 225
90. Correns, C. (1899). *Untersuchungen über die Vermehrung der Laubmoose durch Brutorgane und Stecklinge,* Jena
91. Cronquist, A. (1960). Bot. Rev. **26**, 425
92. Crundwell, A. C. (1960). Trans. Brit. bryol. Soc. **3**, 706
93. Crundwell, A. C. (1962). Trans. Brit. bryol. Soc. **4**, 334
94. Crundwell, A. C. (1970). Biol. J. Linn. Soc. **2**, 221
95. Crundwell, A. C., & Nyholm, E. (1964). Trans. Brit. bryol. Soc. **4**, 597
96. Davis, B. M. (1903). Ann. Bot. **17**, 477
97. Davis, P. H., & Heywood, V. H. (1963). *Principles of Angiosperm Taxonomy,* Edinburgh, 556 pp.
98. de Bergevin, E. (1902). Rev. bryol. **29**, 115
99. Degenkolbe, W. (1937). Annls. bryol. **10**, 43
100. Demaret, F., & Castagne, E. (1959–). *Flore générale de Belgique* vol. 2: *Bryophytes,* in fascicles, Brussels
101. Denizot, J. (1963). Rev. bryol. lichen. **32**, 73
102. Dickson, J. H. (1967). Rev. Palaeobot. Palynol. **2**, 245
103. Dickson, J. H. (1968). Trans. Brit. bryol. Soc. **5**, 588
104. Dickson, J. H. (1969). Nyt Mag. Bot. **16**, 237
105. Diers, L. (1966). J. Cell Biol. **28**, 527
106. Diers, L. (1967). Planta **72**, 119
107. Dixon, H. N. (1924). *The student's handbook of British mosses,* 3rd ed., London, 581 pp.
108. Dixon, H. N. (1932). Classification of mosses. In Verdoorn, F., *Manual of Bryology,* 397
109. Dixon, H. N. (1960). Mosses of Tristan da Cunha. Results of Norwegian Scientific Expedition to Tristan da Cunha 1937–38, no.48, Oslo
110. Doignon, P. (1947). Bryophytes. In *Flore du massif de Fontainebleau,* Paris
111. Doignon, P. (1949). Rev. bryol. lichen. **18**, 160
112. Doignon, P. (1950). Rev. bryol. lichen. **19**, 208
113. Doignon, P. (1952). Rev. bryol. lichen. **21**, 244
114. Duda, J., & Váňa, J. (1969). Čas. slezsk. Mus. Opavě, ser, A. **18**, 21
115. Düll, R. (1969). Herzogia **1**, 215
116. Eschrich, W. J., & Steiner, M. (1967). Planta **74**, 330
117. Eschrich, W. J., & Steiner, M. (1968). Planta **82**, 321
118. Evans, A. W. (1910). Ann. Bot. **24**, 271
119. Evans, A. W. (1912). Ann. Bot. **26**, 1
120. Evans, A. W. (1939). Bot. Rev. **5**, 49

121. Eymé, J., & Suire, C. (1967). C. r. hebd. Séanc. Acad. Sci., Paris, sér. D. 265, 1788
122. Fleischer, M. (1902-3). *Die Musci der Flora von Buitenzorg,* Leiden
123. Fleischer, N. (1920). Hedwigia 61, 390
124. Florschütz, P. A. (1964). *The mosses of Suriname,* pt.1, Leiden, 271 pp.
125. Foote, K. G. (1966). Bryologist, 69, 265
126. Forman, R. T. T. (1964). Ecol. Monogr. 34, 1
127. Fritsch, F. E., & Salisbury, E. J. (1915). New Phytol. 14, 116
128. Fulford, M. (1946). Ann. Cryptog. Phytop. Waltham, Mass. 3, 173 pp.
129. Fulford, M. (1948). Bot. Rev. 14, 127
130. Fulford, M. (1951). Evolution, 5, 243
131. Fulford, M. (1955). Rev. bryol. lichen. 24, 41
132. Fulford, M. (1956). Phytomorphology 6, 199
133. Fulford, M. (1963-8) *Manual of the Leafy Hepaticae of Latin America,* pts.I-III, Mem. N.Y. Bot. Gdn., 392 pp.
134. Fulford, M. (1964). Phytomorphology 14, 103
135. Gams, H. (1932). Bryo-cenology. In Verdoorn, F., *Manual of Bryology,* 323
136. Gams, H. (1953). Rev. bryol. lichen. 22, 161
137. Gams, H. (1959). Rev. bryol. lichen. 28, 326
138. Gaume, R. (1953). Rev. bryol. lichen. 22, 141
139. Gaume, R. (1956). Rev. bryol. lichen. 25, 1
140. Gemmell, A. R. (1950). New Phytol. 49, 64
141. Gilbert, O. L. (1968). New Phytol. 67, 15
142. Gimingham, C. H. (1948). Trans. Brit. bryol. Soc. 1, 70
143. Gimingham, C. H., & Birse, E. M. (1957). J. Ecol. 45, 533
144. Gimingham, C. H., & Robertson, E. T. (1950). Trans. Brit. bryol. Soc. 1, 330
145. Gimingham, C. H., et al. (1961). Trans. Proc. Bot. Soc. Edinb. 39, 125
146. Godwin, H. (1956). *The History of the British Flora,* Cambridge, 384 pp.
147. Godwin, H., & Richards, P. W. (1946). Rev. bryol. lichen. 15, 123
148. Goebel, K. (1905). *Organography of Plants,* vol.2, transl. by Bayley-Balfour, Oxford. There is a later German edition (1930)
149. Gorton, B. S., & Eakin, R. E. (1957). Bot. Gaz. 119, 31
150. Gourgaud, M. (1965). C. r. hebd. Séanc. Acad. Sci. Paris 261, 3185
151. Greene, S. W. (1957). Trans. Brit. bryol. Soc. 3, 181
152. Greene, S. W., & Greene, D. M. (1960). Trans. Brit. bryol. Soc. 3, 715
153. Greig-Smith, P. (1958). Trans. Brit. bryol. Soc. 3, 418
154. Grolle, R. (1963). Öst. bot. Z. 110, 444
155. Grolle, R. (1967). J. Hattori bot. Lab. 30, 1
156. Grønlie, A. M. (1948). Nytt. Mag. Naturvid. 86, 117
157. Grout, A. J. (1928-40). *Moss flora of North America,* 3 vols., Newfane, Vermont
158. Grubb, P. J. (1970). New Phytol. 69, 303
159. Grubb, P. J., Flint, O. P., & Gregory, S. C. (1969). Trans. Brit. bryol. Soc. 5, 802
160. Guillamot, M. (1949). Bull. Soc. bot. Fr. 96, 242
161. Gullvåg, B. M. (1967). J. Palynol. 2, 49
162. Gyorffy, I. (1940). Bull. Inst. roy. Hist. nat. Sofia 13, 207
163. Haberlandt, G. (1914). *Physiological Plant Anatomy* (transl. by Drummond, M., from 4th German ed.), London, 777 pp.
164. Harris, T. M. (1938). *The British Rhaetic Flora,* Brit. Museum (Nat. Hist.), London, 84 pp.
165. Harris, T. M. (1961). *The Yorkshire Jurassic Flora,* vol.1, Brit. Museum (Nat. Hist.), London, 212 pp.
166. Haskell, G. (1949). Bryologist 52, 49

167. Hatcher, R. E. (1960–1). Nova Hedwigia 2, 573; 3, 1
168. Hatcher, R. E. (1967). Brittonia 19, 178
169. Hattori, S., & Inoue, H. (1958). J. Hattori bot, Lab. 19, 133
170. Hattori, S., & Mizutani, M. (1958). J. Hattori bot. Lab. 20, 295
171. Haupt, A. W. (1926). Bot. Gaz. 82, 30
172. Hébant, C. (1966). C. r. hebd. Séanc. Acad. Sci., Paris 262, 2585
173. Hébant, C. (1967). C. r. hebd. Séanc. Acad. Sci. Paris 264, 901
174. Hébant, C. (1967). Naturalia Monspeliensia, Sér. Bot. 18, 293
175. Hébant, C. (1968). Naturalia Monspeliensia, Sér. Bot. 19, 75
176. Hébant, C. (1968–9). Rev. bryol. lichen. 36, 721
177. Heintze, A. (1927). Cormophytetmas Fylogeni (Phylogeny of the cormophytes), Lund, 170 pp.
178. Heitz, E. (1945). Arch. Klaus-Stift. Vererb–Forsch. 20 (Ergänzungsband), 119
179. Hertzfelder, H. (1921). Flora 114, 385
180. Herzog, T. (1926). Geographie der Moose, Jena, 439 pp.
181. Herzog, T. (1942). Flora 136, 264
182. Herzog, T. (1952). Rev. bryol. lichen. 21, 46
183. Herzog, T., & Grolle, R. (1958). Rev. bryol. lichen. 27, 147
184. Hoffman, G. R. (1966). Bryologist 69, 182
185. Hoffman, G. R. (1966). Ecol. Monogr. 36, 157
186. Höfler, K. (1946). Anz. Akad. Wiss. Wien 3, 5
187. Hofmeister, W. (1862). On the germination, development and fructification of the higher Cryptogamia, Ray Society (translation), London
188. Holferty, G. M. (1904). Bot. Gaz. 37, 106
189. Hong, W. S. (1966). Bryologist 69, 393
190. Hope-Simpson, J. F. (1941). J. Ecol. 29, 107
191. Hooker, W. J. (1816). British Jungermanniae, London
192. Horikawa, Y., & Miyoshi, N. (1965). Hikobia 4, 148
193. Hörmann, H. (1959). Nova Hedwigia 1, 203
194. Hosokawa, T., & Kubota, H. (1957). J. Ecol. 45, 579
195. Howe, M. A. (1899). Mem. Torrey bot. Club 7, 1
196. Hueber, F. M. (1961). Ann. Mo. Bot. Gdn. 48, 125
197. Hughes, J. G. (1969). New Phytol. 68, 883
198. Huneck, S. (1969). J. Hattori bot. Lab. 32, 1
199. Husnot, T. (1884–94). Muscologia gallica, 2 vols., Paris, 458 pp.
200. Husnot, T. (1922). Hepaticologia gallica, 2e éd., Paris, 163 pp.
201. Hy, F. (1884). Ann. Sci. nat. bot. VI. 18, 105
202. Ikeno, S. (1903). Bot. Centralbl. 15, 65
203. Ingold, C. T. (1939). Spore discharge in land plants, Oxford, 178 pp.
204. Ingold, C. T. (1959). Trans. Proc. bot. Soc. Edinb. 38, 76
205. Inoue, H. (1959). Bot. Mag. Tokyo 72, 131
206. Inoue, H. (1960). J. Hattori bot. Lab. 23, 148
207. Inoue, H. (1965). Bot. Mag. Tokyo 78, 220
208. Inoue, H. (1967). J. Hattori bot. Lab. 30, 54
209. Inoue, H. (1967). Bot. Mag. Tokyo 80, 172
210. Irmscher, E. (1912). Jb. wiss, Bot. 50, 387
211. Isoviita, P. (1966). Ann. Bot. Fenn. 3, 199
212. Iwatsuki, Z. (1960). J. Hattori bot. Lab. 22, 159
213. Iwatsuki, Z., & Hattori, S. (1965). J. Hattori bot. Lab. 28, 221
214. Iwatsuki, Z., & Hattori, S. (1966). J. Hattori bot. Lab. 29, 223
215. Iwatsuki, Z., & Sharp, A. J. (1967). J. Hattori bot. Lab. 30, 152
216. Jack, J. B. (1895). Flora 81, 1
217. Jennings, E. O. (1928). Bryologist 31, 10

218. Joensson, B., and Olin, E. (1898). Lunds Univ. Arsskr. **34**, 440

219. Johnson, D. S. (1904). Bot. Gaz. **38**, 85

220. Jones, E. W. (1952–3). Trans. Brit. bryol. Soc. **2**, 19, 220

221. Jones, E. W. (1958). Trans. Brit. bryol. Soc. **3**, 353

222. Jones, E. W. (1958). J. W. Afr. Sci. Ass. **4**, 50

223. Jones, E. W. (1962). Trans. Brit. bryol. Soc. **4**, 254

224. Jones, E. W. (1969). Trans. Brit. bryol. Soc. **5**, 775

225. Jovet-Ast, S. (1967). Bryophyta. In Boureau, E., *Traité de Paléobotanique* (Ed. Masson), 19–175

226. Kachroo, P. (1955). J. Hattori bot, Lab. **15**, 70

227. Kamerling, Z. (1898). Flora **85**, 157

228. Kashyap, S. R., & Dutt, N. L. (1925). Proc. Lahore phil. Soc. **4**, 49

229. Kawai, I. (1968). Sci. Report Kanazawa Univ. **13**, 127

230. Kernbach, B. (1964). Z. Bot. **52**, 173

231. Khanna, K. R. (1960). Caryologia **13**, 559

232. Khanna, K. R. (1965). Evolution **18**, 652

233. Khanna, K. R. (1967). Univ. Colo. Stud. ser. biol. **26**, 1

234. Kitagawa, N. (1965). J. Hattori bot, Lab. **28**, 239

235. Klein, B. (1967). Planta **73**, 12

236. Knapp, E. (1930). *Untersuchungen über die Hüllorgane um Archegonien u. Sporogonien der akrogynen Jungermaniaceen,* Bot. Abhandl. (issued by K. Goebel), Jena, 168 pp.

237. Knox, E. M. (1939). Trans, Proc. Bot. Soc. Edinb. **35**, 109

238. Koch, L. F. (1956). Bryologist **59**, 23

239. Kofler, L. (1959). Rev. bryol. lichen. **28**, 1

240. Kraemer, H. (1901). Bot. Gaz. **32**, 422

241. Kreh, W. (1909). Nova Acta Leop.–Carol. Akad. **90**, 89 pp.

242. Kuc, M. (1964). Monogr. bot., Warsaw, 212 pp.

243. Lacey, W. S. (1969). Biol. Rev. **44**, 189

244. Lal, M. (1963). In *Plant Tissue and Organ Culture,* a symposium by Maheshwari and Swamy, 363

245. Land, W. J. G. (1919). Bot. Gaz. **68**, 392

246. Landwehr, J. (1951). Buxbaumia **5**, 26

247. Lang, W. H. (1907). Ann. Bot. **21**. 201

248. Lange, O. L. (1955). Flora **142**, 381

249. Lawton, E. (1965). Bull. Torrey bot. Club **92**, 333

250. Lazarenko, A. S. (1957). Bryologist **60**, 14

251. Lazarenko, A. S. (1964). Dopov. Akad. Nauk ukr. RSR, ser. B. **4**, 541

252. Lazarenko, A. S. (1965). Tsitologia i Genetika 1965, 158

253. Lazarenko, A. S. (1967). Tsitologia i Genetika 1967, 15

254. Leach, W. (1930). J. Ecol. **18**, 324

255. Leblanc. F., & Rao, D. N. (1966). Bryologist **69**, 338

256. Leitgeb, H. (1875). *Untersuchungen über die Lebermoose,* heft 2: *Die Foliosen Jungermannien,* Jena

257. Leitgeb, H. (1879). *Untersuchungen über die Lebermoose,* heft 5: *Die Anthoceroteen,* Graz

258. Leitgeb, H. (1881). *Die Marchantiaceen und allgemeine Bemerkungen über Lebermoose,* Graz

259. Lewis, D. H. (1970). Trans. Brit. bryol. Soc. **6**, 108

260. Lewis, K. R. (1961). Trans. Brit. bryol. Soc. **4**, 111

261. Limpricht, K. G. (1890–1904). *Die Laubmoose Deutschlands, Österreichs und der Schweiz.* Rabenhorst's *Kryptogamen-flora,* 3 vols., Leipzig

262. Lodge, E. (1959). J. Linn, Soc. Lond. **56**, 218

263. Longton, R. E. (1962). Trans. Brit. bryol. Soc. **4**, 326

264. Longton, R. E., & Greene, S. W. (1967). Phil. Trans. R. Soc. Lond. B. **252**, 295

265. Longton, R. E., & Greene, S. W. (1969). Ann. Bot. **33**, 107

266. Lorbeer, G. (1934). Jb. wiss. Bot. **80**, 565

267. Lorch, W. (1931). Anatomie der Laubmoose. In Linsbauer, K., *Handbuch der Pflanzenanatomie,* bd. VII, Berlin

268. Lotsy, J. P. (1909). *Vorträge über botanische Stammesgeschichte,* Jena

269. Lowry, R. J. (1948). Mem. Torrey bot. Cl. **20**, 1

270. Lowry, R. J. (1954). Bryologist **57**, 1

271. Lundblad, B. (1954). Svensk bot. Tidskr. **48**, 381

272. Lye, K. A. (1970). Nytt. Mag. Bot. **17**, 25

273. McCleary, J. A., & Walkington, D. L. (1966). Rev. bryol. lichen. **34**, 309

274. McClure, J. W., & Miller, H. A. (1967). Nova Hedwigia **14**, 111

275. McClymont, J. W., & Larson, D. A. (1964). Am. J. Bot. **51**, 195

276. MacQuarrie, G., & Maltzahn, K. E. V. (1959). Can. J. Bot. **37**, 121

277. Macvicar, S. M. (1926). *The student's handbook of British Hepatics,* London, 464 pp.

278. Mägdefrau, K. (1935). Z. Bot. **29**, 337

279. Magnée, C. (1967). Naturalistes belg. **48**, 325

280. Mahabele, T. S. (1942). Proc. Indian Acad. Sci., ser. B **16**, 141

281. Malta, N. (1921). Acta Univ. Latviensis **1**, 125

282. Manton, I. (1957). J. exper. Bot. **8**, 382

283. Manton, I., & Clarke, B. (1952). J. exper. Bot. **3**, 265

284. Martin, W. (1951). Trans. Brit. bryol. Soc. **1**, 471

285. Meeuse, A. D. J. (1966). Acta bot. néerl. **15**, 162

286. Mehra, P. N. (1957). Amer. J. Bot. **44**, 505, 573

287. Mehra, P. N., & Kachroo, P. (1951). Bryologist **54**, 1

288. Menge, F. (1930). Flora **124**, 423

289. Meusel, H. (1935). Not. Act. Leopold. N.F. **3**, 123

290. Mirimanoff-Olivet, A. (1943). Ber. Schweiz. bot. Ges. **53**, 389

291. Mitra, G. C., & Allsopp, A. (1959). Nature (Lond.) **183**, 974

292. Mitra, G. C., Allsopp, A., & Wareing, P. F. (1959). Phytomorphology **9**, 47

293. Mitra, G. C., Misra, L. P., & Prabha, C. (1965). Planta **65**, 42

294. Miyoshi, N. (1967). Bull. Okayama Coll. Sci. **3**, 35

295. Miyoshi, N. (1969). Hikobia **5**, 172

296. Mizutani, M. (1966). J. Hattori bot. Lab. **29**, 153

297. Monkemeyer, W. (1927). *Die Laubmoose Europas,* IV: *Erganzungsband,* Leipzig, 960 pp.

298. Muggoch, H., & Walton, J. (1942). Proc. roy. Soc. B. **130**, 448

299. Müller, K. (1940). Hedwigia **79**, 72

300. Müller, K. (1954). *Die Lebermoose Europas. Rabenhorst's Kryptogamen-Flora,* bd. 6, 1, Leipzig. 756 pp.

301. Nehira, K. (1962). Hikobia **3**, 96

302. Nehira, K. (1966). J. Sci. Hiroshima Univ., ser. B, divn. 2, **11**, 1

303. Nehira, K. (1967). Hikobia **5**, 39

304. Neuburg, M. F. (1956). C. R. Acad. Sci. USSR **107**, 2

305. Neuburg, M. F. (1958). J. palaeontolog. Soc. Ind. (Sahni Mem. no.) **3**, 23

306. Neuburg, M. F. (1960). Akad. Nauk USSR **19**, 104 pp.

307. Noguchi, A. (1948–51). J. Hattori bot. Lab. **3**, 53; **4**, 1; **5**, 7

308. Noguchi, A., and Osada, T. (1960). J. Hattori bot. Lab. **23**, 122

309. Nyholm, E. (1954–69). *Moss Flora of Fennoscandia,* II: *Musci,* 6 fascicles, Lund, 799 pp.

310. Ochi, H. (1957). Jap. J. Ecol. **7**, 51
311. Ochi, H. (1959). Biolog. Inst. Tottori., 124 pp.
312. Oehlkers, F. (1965). Z. Vererbungslehre **96**, 234
313. O'Hanlon, Sr. M. E. (1926). Bot. Gaz. **82**, 215
314. Osada, T. (1965-6). J. Hattori bot. Lab. **28**, 171; **29**, 1
315. Pagan, F. M. (1932). Bot. Gaz. **93**, 71
316. Pais, M. S. (1966). Revta Biol. Lisb. **5**, 239
317. Pandé, S. K. (1932). J. Indian bot. Soc. **11**, 169
318. Pandé, S. K. (1934). Proc. Indian Acad. Sci., B. **5**, 205
319. Paolillo, D. J. Jr. (1965). Can. J. Bot. **43**, 669
320. Parihar, N. S. (1965). *An introduction to Embryophyta*, vol.1: *Bryophyta*, 5th ed., Allahabad, 377 pp.
321. Paton, J. A. (1954). Trans. Brit. bryol. Soc. **2**, 349
322. Paton, J. A. (1956). Trans. Brit. bryol. Soc. **3**, 103
323. Paton, J. A. (1961). Trans. Brit. bryol. Soc. **4**, 1
324. Paton, J. A., & Goodman, P. J. (1955). Trans. Brit. bryol. Soc. **2**, 561
325. Paton, J. A., & Newton, M. E. (1967). Trans. Brit. bryol. Soc. **5**, 226
326. Paton, J. A., & Pearce, J. V. (1957). Trans. Brit. bryol. Soc. **3**, 228
327. Peirce, G. J. (1906). Bot. Gaz. **42**, 55
328. Perring, F. (1959). J. Ecol. **47**, 447
329. Perring, F. (1960). J. Ecol. **48**, 415
330. Perry, A. R., & Warburg, E. F. (1962). Trans. Brit. bryol. Soc. **4**, 335
331. Persson, H. (1954). Bot. Notiser 1954, 39
332. Persson, H. (1956). J. Hattori bot. Lab. **17**, 1
333. Persson, H. (1966). Miscnea bryol. lichen. **4**, 25
334. Pettet, A. (1964). Trans. Brit. bryol. Soc. **4**, 642
335. Philibert, H. (1884). Rev. bryol. **11**, 49 (and other papers)
336. Pócs, T. (1965). Az. Egri Tanárképzö Foisköla Füzetei no. 371, 453
337. Poliakov, I. A., Leontév, A. M., & Mel'nikov, L. K. (1962). Pochvovedinie **11**, 45
338. Pringsheim, N. (1878). Über Sprossung der Moosfrüchte und den Generations-wechsel der Thallophyten. Jahrb. Wiss. Bot. **11**, 1
339. Proctor, M. C. F. (1956). Trans. Brit. bryol. Soc. **3**, 1
340. Proctor, M. C. F. (1964). In *Dartmoor Essays*, p. 141, Torquay
341. Proctor, M. C. F. (1967). J. Ecol. **55**, 119
342. Proctor, V. W. (1961). Bryologist **64**, 58
343. Proskauer, J. (1948). Ann. Bot. N.S. **12**, 237, 427
344. Proskauer, J. (1951). Bryologist **54**, 243
345. Proskauer, J. (1954). J. Linn. Soc. Bot. **55**, 143
346. Proskauer, J. (1960). Phytomorphology **10**, 1
347. Proskauer, J. (1961). Taxon **10**, 155
348. Proskauer, J. (1961). Phytomorphology **11**, 359
349. Ramsay, H. P. (1964). Bryologist **67**, 153
350. Rancken, H. (1914). Acta Soc. Fauna Flora fenn. **39**, no.2 (Dissertation)
351. Rao, D. N., & Leblanc, F. (1967). Bryologist 70, 141
352. Ratcliffe, D. A. (1968). New Phytol. **67**, 365
353. Ratcliffe, D. A., & Walker, D. (1958). J. Ecol. **46**, 407
354. Reed, C. F., & Robinson, H. (1967). Phytologia **15**, 61
355. Reese, W. D. (1961). Bryologist **64**, 89
356. Reese, W. D., & Lemmon, B. E. (1965). Bryologist **68**, 277
357. Reimers, H. (1954). Bryophyta. In Engler, *Syllabus der Pflanzenfamilien*, 218–68
358. Reissinger, A. (1950). Palaeontographica **90B**, 99
359. Richards, P. W. (1929). J. Ecol. **17**, 127

360. Richards, P. W. (1932). Ecology. In Verdoorn, F., *Manual of Bryology,* 367
361. Richards, P. W. (1947). Trans. Brit. bryol. Soc. 1, 1
362. Richards, P. W. (1959). Bryophyta. In *Vistas of Botany* (Ed. Turrill, W. B.), 387
363. Richards, P. W., & Wallace, E. C. (1950). Trans. Brit. bryol. Soc. 1, 427
364. Ridgway, J. E. (1967). Ann. Mo. bot. Gdn. 54, 95
365. Robinson, H. (1967). Bryologist 70, 1
366. Romell, L. G. (1939;. Svensk bot. Tidskr. 33, 366
367. Rose, F. (1949–51). Trans. Brit. bryol. Soc. 1, 202, 255, 427
368. Rose, F. (1953). Proc. Linn. Soc. Lond. 164, 186
369. Roth, G. (1904–5). *Die europaischen Laubmoose,* 2 vols., Leipzig, 1331 pp.
370. Ruhland, W. (1924). Musci. Allgemeiner Teil. In Engler and Prantl, *Die natürlichen Pflanzenfamilien,* Bd. 10, Leipzig
371. Rühling, A., & Tyler, G. (1968). Bot. Notiser 121, 321
372. Sainsbury, G. O. K. (1955). *A handbook of the New Zealand mosses,* Wellington, N.Z., 490 pp.
373. Sato, S. (1956). Bot. Mag. Tokyo 69, 435
374. Savicz-Ljubitskaja, L. I., & Abramov, I. I. (1959). Rev. bryol. lichen. 28, 330
375. Schelpe, E. A. C. L. E. (1969). J. S. Afr. Bot. 35, 109
376. Schiffner, V. (1893). Hepaticae. In Engler and Prantl, *Die natürlichen Pflanzenfamilien,* 1 Th., 3 Abt.
377. Schiffner, V. (1913). Öst. bot. Z. 63
378. Schofield, W. B. (1965). J. Hattori bot. Lab. 28, 17
379. Schostakowitsch, W. (1894). Flora 79, 350
380. Schultze-Motel, W. (1963). Willdenowia 3, 399
381. Schuster, R. M. (1951). Amer. Midl, Nat. 45, 1
382. Schuster, R. M. (1962). Trans. Brit. bryol. Soc. 4, 230
383. Schuster, R. M. (1964). J. Hattori bot. Lab. 27, 183
384. Schuster, R. M. (1966). *Hepaticae and Anthocerotae of North America,* vol. 1, New York, 802 pp.
385. Schuster, R. M. (1966). Nova Hedwigia 13, 1
386. Schuster, R. M. (1969). Taxon 18, 46
387. Schuster, R. M., & Hattori, S. (1954). H. Hattori bot. Lab. 11, 11
388. Schuster, R. M., & Scott, G. A. M. (1969). J. Hattori bot. Lab. 32, 219
389. Schwarz, O. (1955). Mitt. Thüring bot. Gesell. 1, 292
390. Scott, D. H. (rev. by Ingold, C. T., 1955). *Introduction to Structural botany,* 1: *Flowerless plants,* London
391. Scott, G. A. M. (1965). J. Ecol. 53, 21
392. Segawa, M. (1965). J. Sci. Hiroshima Univ. 10, 69
393. Shacklette, H. T. (1961). Bryologist 64, 1
394. Shacklette, H. T. (1967). Bull. U.S. geol. Surv. 1198, G, 18
395. Shin, T. (1964). Scient. Rep. Kagoshima Univ. 13, 35
396. Showalter, A. M. (1926). Ann. Bot. 40, 961
397. Showalter, A. M. (1928). Cellule 38, 295
398. Siegel, S. M. (1969). Am. J. Bot. 56, 175
399. Sinoir, Y. (1952). Rev. bryol. lichen. 21, 32
400. Sironval, C. (1947). Bull. Soc. bot. Belg. 79, 48
401. Skene, M. (1915). Ann. Bot. 29, 65
402. Smith, A. J. E., & Newton, M. E. (1966–8). Trans. Brit. bryol. Soc. 5, 117, 245, 463
403. Smith, G. M. (1955). *Cryptogamic botany,* vol. 2. (2nd ed.), London
404. Smith, J. L. (1966). Univ. of California publications in Botany, vol. 39, 1
405. Sowter, F. A. (1948). Trans. Brit. bryol. Soc. 1, 73
406. Springer, E. (1935). Zeitschr, f. Vererbungslehre 69, 249

407. Spruce, R. (1884). Trans. bot. Soc. Edinb. **15**, 1

408. Stange, L. (1964). Advances in Morphogenesis **4**, 111

409. Steere, W. C. (1947). Bryologist **50**, 247

410. Steere, W. C. (1954). Bot. Gaz. **116**, 93

411. Størmer, P. (1969). *Mosses with a Western and Southern distribution in Norway*, Oslo, 288 pp.

412. Streeter, D. T. (1965). Trans. Brit. bryol. Soc. **4**, 818

413. Streeter, D. T. (1970). Sci. Prog., Oxf. **58**, 419

414. Strunk, R. (1914). *Beiträge zur Kenntnis der Organisation der Moose*, Dissert., Bonn

415. Svensson, G. (1965). Bot. Notiser 1965, 49

416. Swinscow, T. D. V. (1959). Trans. Brit. bryol. Soc. **3**, 509

417. Szafran, B. (1963). *Bryophyta*, 1, Polska Akad. Nauk Inst. bot., Warsaw

418. Takaki, N. (1964). J. Hattori bot. Lab. **27**, 73

419. Tallis, J. H. (1958–9). J. Ecol. **46**, 271; **47**, 325

420. Tallis, J. H. (1964). J. Ecol. **52**, 345

421. Tamm, C. O. (1953). Medd. Skogsforskn.–Inst. Stockholm **43**, 140 pp.

422. Tansley, A. G. (1939). *The British Islands and their Vegetation*, Cambridge

423. Tansley, A. G., & Chick, E. (1901). Ann. Bot. **15**, 1

424. Tatuno, S. (1955). J. Hattori bot. Lab. **14**, 109

425. Tatuno, S. (1958). J. Hattori bot. Lab. **20**, 119

426. Taylor, E. C., Sr. (1962). Bryologist **65**, 175

427. Townrow, J. (1959). J. S. Afr. Bot. **25**, 1

428. Udar, R., & Chandia, V. (1964). Bryologist **67**, 55

429. Vaarama, A. (1956). Irish Nat. J. **12**, 30

430. van Andel, O. M. (1952). Trans. Brit. bryol. Soc. **2**, 74

431. van der Wijk, R., Margadant, W. D., & Florschütz, P. A. (1959–69). *Index Muscorum*, 5 vols., Utrecht, 3138 pp.

432. van Zanten, B. O. (1964). Nova Guinea, Botany **16**, 263

433. Verdoorn, F. (1950). Bryologist **53**, 1

434. Visotska, E. I. (1967). Tsitologia i Genetika **1**, 30

435. Vöchting, H. (1886). Jb. wiss. Bot. **16**, 367

436. Voth, P. D., & Hamner, K. C. (1940). Bot. Gaz. **102**, 169

437. Walker, R., & Pennington, W. (1939). New Phytol. **38**, 62

438. Wallace, E. C. (1950). Trans. Brit. bryol. Soc. **1**, 327

439. Walton, J. (1925). Ann. Bot. **39**, 563

440. Walton, J. (1928). Ann. Bot. **42**, 707

441. Walton, J. (1943). Nature (Lond.) **152**, 51

442. Warburg, E. F. (1949). Trans. Brit. bryol. Soc. **1**, 199

443. Watson, E. V. (1950). Trans. Brit. bryol. Soc. **1**, 345

444. Watson, E. V. (1953). Trans. Proc. bot. Soc. Edinb. **36**, 165

445. Watson, E. V. (1957). Famous plants, 6: Funaria, New Biology, **22**, 104

446. Watson, E. V. (1960). J. Ecol. **48**, 397

447. Watson, E. V. (1960). Trans. Proc. bot. Soc. Edinb. **39**, 85

448. Watson, E. V. (1968). *British Mosses and liverworts* (2nd ed.), Cambridge, 495 pp.

449. Watson, H. (1947). *Woodland mosses,* Forestry Commission booklet no.1, London, H.M. Stationery Office

450. Watson, W. (1913). New Phytol. **13**, 149

451. Watson, W. (1918). J. Ecol. **6**, 126, 189

452. Watson, W. (1925). J. Ecol. **13**, 22

453. Watson, W. (1932). J. Ecol. **20**, 284

454. Watt, A. S., & Jones, E. W. (1948). J. Ecol. **36**, 283

455. Welch, W. H. (1960). *A monograph of the Fontinalaceae,* The Hague, 357 pp.

456. Welch, W. H. (1962). Bryologist **65**, 1
457. Wettstein, F. von (1923). Biol. Zbl. **43**, 71
458. Wettstein, F. von (1924). Z. indukt. Abstamm. Vererbl. **33**, 1
459. Wettstein, F. von (1940). Ber. dtsch. bot. Ges. **58**, 374
460. Wettstein, F. von, and Straub, J. (1942). Z. indukt. Abstamm. Vererbl. **80**, 271
461. Wettstein, R. (1924). *Handbuch der systematischen Botanik, Aufl.* 3, Leipzig, 1017 pp.
462. Whitehouse, H. L. K. (1961). Trans. Brit. bryol. Soc. **4**, 84
463. Whitehouse, H. L. K. (1966). Trans. Brit. bryol. Soc. **5**, 103
464. Wigglesworth, G. (1947). Trans. Brit. bryol. Soc. **1**, 4
465. Williams, S. (1950). Trans. Brit. bryol. Soc. **1**, 357
466. Willis, A. J. (1964). Trans. Brit. bryol. Soc. **4**, 668
467. Wilson, M. (1908). Ann. Bot. **22**, 328
468. Wilson, M. (1911). Ann. Bot. **25**, 415
469. Wilson, W. (1855). *Bryologia britannica,* London, 445 pp.
470. Winkler, S. (1967). Rev. bryol. lichen. **35**, 303
471. Wylie, A. P. (1957). Trans. Brit. bryol. Soc. **3**, 260
472. Yano, K. (1956). Bot. Mag. Tokyo **69**, 156
473. Yano, K. (1957). Mem. Fac. Ed. Niigata Univ. **6**, no.3, 1
474. Yarranton, G. A. (1962). Rev. bryol. lichen. **31**, 168
475. Yarranton, G. A. (1967). Can. J. Bot. **45**, 93
476. Zacherl, H. (1956). Z. Bot. **44**, 409
477. Zielinski, F. (1910). Flora **100**, 1

INDEX